Daniel Gentner

Palm Theory, Mass Transports and Ergodic Theory for Group-Stationary Processes

Palm Theory, Mass Transports and Ergodic Theory for Group-Stationary Processes

by
Daniel Gentner

Dissertation, Karlsruher Institut für Technologie
Fakultät für Mathematik
Tag der mündlichen Prüfung: 16. Februar 2011

Impressum

Karlsruher Institut für Technologie (KIT)
KIT Scientific Publishing
Straße am Forum 2
D-76131 Karlsruhe
www.ksp.kit.edu

KIT – Universität des Landes Baden-Württemberg und nationales
Forschungszentrum in der Helmholtz-Gemeinschaft

KIT Scientific Publishing 2011
Print on Demand

ISBN 978-3-86644-669-4

Gratefully dedicated to my parents
Claudia and Hermann

and to my dear sister
Anja.

PREFACE

This PhD thesis has been written during my employment as research and teaching assistant at the Institute for Stochastics at Karlsruhe Institute of Technology (KIT). I wish to thank several people at this place for various kinds of support.

Special thanks go to my advisor Prof. Dr. Günter Last for introducing me to the realms of random measures and Stochastic Geometry, for valuable input and hints for my research throughout these three years and his constant readiness to discuss my math problems. My second reviewer, apl. Prof. Dr. Daniel Hug, not only found a number of typing errors in a first version of this text (the remaining errors are all mine) but also provided valuable comments and help, especially concerning questions from convex geometry. I also benefited from discussions with PD Dr. Dieter Kadelka, PD Dr. Stefan Kühnlein, HDoz. Dr. Oliver Baues, Dr. Sebastian Grensing and Dr. Panayotis Mertikopoulos. Dr. Martin Folkers regularly called my attention to new books on several different areas of mathematics and I wish to thank him for creating and maintaining an excellent mathematical library from which I benefited regularly. I would like to thank all members of the Institute of Stochastics, in particular Michaela Regelin and Tatjana Dominic, for creating such a friendly and pleasant working environment.

My sincere thanks go to Prof. Dr. Olav Kallenberg who taught me many things through his extraordinary books and through several papers that are intimately linked to parts of this thesis. Equally, I would like to thank Prof. Dr. Panamalai Ramarao Parthasarathy from IIT Madras. He taught me queueing theory when I had the pleasure to TA some of the courses he gave in Karlsruhe and provided constant encouragement and support.

I owe huge thanks to my friends in Germany, Greece and the US as well as to my girlfriend. The year at Brown University in Providence, RI, truly enriched my life both personally and scientifically and I gratefully memorize Brown as an awesome place to learn and do mathematics.

Finally, I am deeply indebted to my family, in particular to my parents and to my sister. Without their recognition, their support of my studies and their encouragement throughout years the writing of this thesis would have been impossible.

Daniel Gentner   Karlsruhe, February 2011

Contents

Chapter 1

Introduction and Outline

In the focus of this thesis is the analysis of *group stationary spatial stochastic processes*. These processes may conveniently be described as group stationary random measures on an appropriate space, where *group stationarity* refers to a distributional invariance property with respect to a group acting on this space. This property represents a requirement, which is weak enough to allow the construction of reasonable and well-fitting models for a large variety of natural real-world phenomena such as e.g. cell growth processes, the development of forest fires, rain-drop models, the arrival and handling of phone calls in a call center or the development of cracks in certain materials. At the same time, it represents a requirement strong enough to allow a reasonable mathematical investigation of these models. Here, a *random measure* is nothing but a random element in the space of all measures on a certain fixed space and this rich class of objects will be of major interest to us (introductions to the subject may be found in [14, 15, 28, 29]). Random measures are in the focus of many researchers since the 1950s and the field developed rapidly since then. It all started with the inspection of integer valued random measures (so called *point processes*) on the real line as a model for the arriving phone calls in a call center. Conrad Palm, a Swedish engineer, was the first who investigated such a model. A central object for the examination of stationary random measures has been named after him - the *Palm measure*. It has been defined so far for random measures living on a space on which a certain group acts *transitively* and with respect to which the measure is stationary. The transitivity together with the stationarity enforce a *complete* statistical spatial homogeneity of the process. Loosely speaking, no matter in which point of space an observer decides to measure the random mass configuration around him, he will never be able to tell from his (repeated) measurements where he sits (no matter which arbitrarily sophisticated statistical toolbox he employs).

The analysis of random measures on $\mathbb{R}^d$ that are stationary with respect to the group of all translations (i.e. with respect to $\mathbb{R}^d$ itself) is by now a classical domain of random measure theory and became an indispensable pillar for the realm of *Stochastic Geometry* (see [64, 66] for comprehensive introductions). Stationary particle processes, k-flat processes, cluster processes, random partitions or tessellations on $\mathbb{R}^d$ have under the stationarity assumption a spatial homogeneity property. This property allows in spite of the fact that they consist realization wise of a discrete set of infinitely many objects (which makes it impossible to average over these objects in the naive sense in such a way that each object receives the same weight), the extraction of meaningful distributions associated to the collection of objects the process consists of. A random element that is distributed according to such a distribution is interpreted as a *typical* object of the relevant process. For the above processes, one speaks of the distribution of a *typical particle*, of a *typical flat*, a *typical cluster* or a *typical cell*. The derivation of all these distributions is the result of a more sophisticated fair averaging procedure, both over the underlying probability space and over the space on which the process lives.

All these distributions may be interpreted as distributions of suitable random objects under the Palm measure, which explains the central role of this measure. It allows an elegant and unified treatment of all the above mentioned distributions. Mecke was the first who used Palm methods in Stochastic Geometry in his seminal paper [47]. Kallenberg contributed in [30] important results on which some of the central results in this thesis are based and that are already published in [21]. We have marked the relevant theorems. Some of these were also independently found by Kallenberg in [31].

This brings us to the first main result of this thesis, namely the derivation of such a Palm measure even for processes (living on some abstract space, possibly different from $\mathbb{R}^d$) which are stationary with respect to arbitrary, possibly non-transitive, group actions. The operating group needs not even be unimodular. As mentioned above, this generalized Palm measure (which we shall call for good reasons the *cumulative Palm measure*) will help us to identify typical objects even for processes, that are not completely stationary, i.e. where the underlying group action is not transitive. This is not only interesting from a theoretical viewpoint. Many real-world phenomena exhibit spatial inhomogeneities, such that the assumption of a complete spatial homogeneity is untenable. Our cumulative Palm measure will allow the mathematical exploration of more adequate, non-transitive models. The reader may think of e.g. a material consisting of different layers with different properties.

A second main part of this thesis is devoted to the extension of an important deterministic principle, the *mass-transport principle*. It constitutes a mass-conservation law for certain transportation rules. Again, this principle has been found for special transitive group actions [24, 6, 7] and has been substantially extended in our paper [21] to the possibly non-transitive and possibly non-unimodular case. As we shall show, it may be interpreted as an identity between cumulative Palm measures which establishes the link to random measure theory, and contributes a valuable intuitive understanding for the transformation of one cumulative Palm measure into another one.

The third main tool that we shall develop for two special types of group stationary processes where the underlying operation in non-transitive, are ergodic theorems. We note here, that Meijering [49] seems to be the first who investigated a random geometric model under ergodicity assumptions. The two classes are, first, random measures on $\mathbb{R}^d$ that are stationary with respect to the operation of a discrete grid, identified by $\mathbb{Z}^d$, and second the case where we have stationarity with respect to the action of a fixed lower dimensional linear subspace of $\mathbb{R}^d$. As it will turn out, the cumulative Palm measure naturally arises under ergodicity assumptions in the limit of certain a.s. and L^p-convergence results, which again shows the relevance of the cumulative Palm measure and establishes the link to Palm theory.

We finally show how this extended toolbox may be used for the inspection of group stationary random processes, where the underlying group action is not necessarily transitive. This includes a result on graph automorphism-stationary random subgraphs of quasi-transitive, possibly non-unimodular deterministic graphs (e.g. a bond percolation model) as well as random partitions on orientable Riemannian manifolds stationary with respect to the natural action of the respective isometry group. It also includes the identification and structural analysis of *typical Cox Delaunay cells*. A Cox-Delaunay tessellation is a special random tessellation of $\mathbb{R}^d$, where the randomness stems from a Cox point process (which is a randomized Poisson process).

The thesis is organized as follows: In Chapter 2 we provide the reader with necessary background in measure theory and also present some recent developments concerning the existence of invariant disintegrations. This chapter already contains some new results. Most notably, we derive a technically elaborate result on the existence of invariant kernels disintegrating any kernel from some measurable space to a product space, exhibiting a certain invariance property.

We then proceed in Chapter 3 with the construction of an important kernel, which is associated to any 'well-behaving' group action. This *inversion kernel* will lead us to the construction of the 'general' Palm measure for arbitrary group actions, the *cumulative Palm measure*. It is derived by factoring out the Haar measure of the operating group from another measure which is naturally associated to the random measure of interest. This cumulative Palm measure is an interesting object of study on its own right, and we shall derive some important theorems around this measure in Chapter 4.

Chapter 5 contains our extension of the mass-transport principle to non-transitive group actions, as well as an important consequence for random measures and transports. In particular, we show that there is a close link between a *transport theorem* for cumulative Palm measures and a version of the mass-transport principle for random transports. We also give a first application of this principle to automorphism-stationary subgraphs of a quasi-transitive, possibly non-unimodular graph.

Our convergence results are the content of Chapter 6. As mentioned above we treat grid-stationarity and subspace-stationarity separately, and we show in the end of that chapter how the cumulative Palm measure naturally arises in the limit under ergodicity assumptions.

The final Chapter 7 will illustrate the usefulness of our developed tools. We give structurally quite explicit formulas for the distribution of typical Cox-Delaunay cells, where the underlying Cox process is assumed to be stationary with respect to a subspace of $\mathbb{R}^d$. This clearly includes the completely stationary case, where the subspace is $\mathbb{R}^d$ itself, and even this special case seems to be new. We also introduce random partitions on Riemannian manifolds and suitably define the notion of *typical cell* and *0-cell*. We derive a relation between these objects which may be paraphrased by the intuitively appealing statement that the 0-cell is a volume-weighted version of the typical cell, a fact well-known for completely stationary tessellations in $\mathbb{R}^d$. We also illustrate the use of our ergodic theorems by deriving some quite general convergence results for grid- or subspace-stationary random tessellations of $\mathbb{R}^d$ and we illustrate these by giving some examples. These examples include for instance $\mathbb{R}$-stationary tessellations on the infinite cylinder $\mathbb{R} \times S^1$ (where the action of $\mathbb{R}$ on $\mathbb{R} \times S^1$ is understood to affect the first component only via translation).

More detailed references to relevant literature may be found in the introductions to each chapter as well as throughout the thesis.

Chapter 2

Fundamentals and recent developments in measure theory

In this introductory chapter we try to familiarize the reader with some basic concepts used in this thesis. These are some elementary as well as some more advanced notions of measure theory in Section 2.1, the notion of Haar measure and group invariant measures in Section 2.2, some recent developments in the theory of invariant disintegrations of jointly invariant measures on product spaces in Section 2.3 and finally the concept of random measures in Section 2.4. Along the lines we also present some new results. These include in Section 2.3 the existence of invariant disintegrations of kernels where each kernel member itself is a jointly invariant measure on a product space. This was established by Gentner and Last [21]. In Chapter 3 this result will be a key ingredient in our existence proof of the *inversion kernel* for even non-transitive group operations, recently independently constructed by Gentner and Last [21] and Kallenberg [31].

2.1 Some notions from measure theory

We shall mainly recall the concepts of universal measurability and kernels in this section. We also fix our basic notation used throughout this thesis. S, T and R shall always denote measurable spaces with respective σ-algebras $\mathcal{S}$, $\mathcal{T}$ and $\mathcal{R}$. Given two σ-algebras $\mathcal{S}$ and $\mathcal{T}$ their product σ-algebra will always be denoted by $\mathcal{S} \otimes \mathcal{T}$. For a measure ν on S and a measurable function $f : S \to [-\infty, \infty]$ we denote the integral $\int f d\nu$ by $\nu f \equiv \nu(f)$ whenever it is well-defined. Further, whenever $(S, \mathcal{S})$ is a measurable space we denote by $\mathcal{S}_+$ the space of $\mathcal{S}$-measurable $[0, \infty]$-valued functions on S. For $f \in \mathcal{S}_+$ we often write $f \cdot \nu$ for the measure $A \mapsto \nu(\mathbf{1}_A \cdot f)$ on $\mathcal{S}$. The power set of a set S is denoted by $\mathcal{P}(S)$ and given a system of subsets $\mathcal{E} \subset \mathcal{P}(S)$ of S the smallest σ-algebra containing $\mathcal{E}$ will be denoted by $\sigma(\mathcal{E})$. If μ and ν are measures on S the relation $\mu \ll \nu$ means that μ is absolutely continuous with respect to ν. The relation $\sim$ denotes mutual absolute continuity between measures and evidently represents an equivalence relation on the space of all measures on a given space. Given a probability space $(\Omega, \mathcal{A}, \mathbb{P})$ the distribution of a random element τ in a measurable space S (also called the *law* of τ) is denoted by $\mathcal{L}(\tau)$.

2.1.1   Regularity properties and u-measurability

In many situations one has to require a measure μ to be σ-*finite*, i.e. to ask for the existence of a measurable partition $B_1, B_2, \ldots$ that splits it into finite pieces, or at least to require it to be *s-finite*, relaxing the above condition by only requiring the existence of a sequence (μ_n) of finite measures that approximates it setwise from below such that $\mu_n(A) \uparrow \mu(A), A \in \mathcal{S}$.

The concept of s-finiteness was used by Kallenberg [30] who noted that it simplifies many arguments, mostly since the class of s-finite measures is closed under projections: an s-finite measure M on a product $S \times T$ induces s-finite measures $M(\cdot \times T)$ and $M(S \times \cdot)$. Note that the analogue statement with s-finiteness replaced by σ-finiteness is wrong: the 2-dimensional Lebesgue measure λ^2 on $\mathbb{R}^2$ is σ-finite, unlike its projection $\lambda^2(\cdot \times \mathbb{R})$. In addition s-finiteness is not only preserved under most basic operations that also preserve σ-finiteness - it is sometimes even easier to verify than a possible σ-finiteness property. Finally most computational rules such as Fubini's Theorem are still valid for s-finite measures.

There are many reasons that make both of these regularity concepts necessary but two particularly important ones are the following.

First, a measure $\mu \neq 0$ on some space S is σ-finite if and only if there is a strictly positive function $f : S \to (0, \infty)$ such that $\mu f < \infty$. This gives rise to an in the sense of mutual absolute continuity *equivalent* finite measure $\nu := f \cdot \mu$ which may clearly be assumed to be a probability measure. Then most spaces possess topological or measure theoretical properties that allow us to identify them with Borel subsets of the reals $\mathbb{R}$. Such spaces are called Borel spaces. For a precise definition, a bijective map $f : S \to T$ between measurable spaces S and T is called a *Borel isomorphism* if both f and f^{-1} are measurable (i.e. the Borel isomorphisms are the isomorphisms in the category of measurable spaces). Now a *Borel space* is a measurable space Borel isomorphic to a Borel subset of $\mathbb{R}$. Hence σ-finite measures on Borel spaces may for some purposes be treated as probability measures on $\mathbb{R}$, where we may use their distribution functions to investigate their properties. As a particularly important example where this observation bears fruits we mention Kallenberg's [30] elegant proof concerning the existence of disintegrations for σ-finite measures on product spaces where the second factor is Borel. It will be essential for our discussions in Subsection 2.3.2.

Second, integration of universally measurable functions that are not measurable with respect to the given σ-algebra is meaningful only if the integrating measure can be approximated by finite measures. Let us quickly repeat the concept of universally measurable sets and functions here. If μ is a measure on $(S, \mathcal{S})$ then we denote by $\mathcal{S}^\mu$ the completion of $\mathcal{S}$ with respect to μ. The *universal completion* of a σ-algebra $\mathcal{S}$ is then defined as

$$\mathcal{S}^u = \bigcap_\mu \mathcal{S}^\mu$$

where the intersection is taken over the class of all probability measures on $(S, \mathcal{S})$ (clearly one may equivalently take the class of all finite measures on $(S, \mathcal{S})$ here). The elements of this σ-algebra are called *universally measurable* sets and a map $f : S \to T$ is called *universally measurable* (*u-measurable* in short) if it is $\mathcal{S}^u/\mathcal{T}$-measurable. Noting that we may decompose any set $A \in \mathcal{S}^u$ for an arbitrary finite measure μ into $A = B \cup C$ with $B \in \mathcal{S}$ and $C \subset N \in \mathcal{S}$ with $\mu(N) = 0$, it is natural to define $\mu(A) := \mu(B)$. This way $\mu(A)$ is clearly well-defined and thus this definition yields a natural extention of μ to $\mathcal{S}^u$. In a next step, we may approximate a given u-measurable function $f : \mathcal{S} \to [0, \infty]$ by u-measurable step functions which defines the integral μf in the obvious way. From integration with respect to finite measures to σ-finite or even more generally s-finite measures is a small step: just note that any such measure μ may be written as a countable sum of finite measures which yields the desired extension.

Recall that a measurable space is *Borel* by definition if it is isomorphic to a Borel subset of $\mathbb{R}$ in the category of measurable spaces, where isomorphisms are bijections between measurable spaces that are measurable in both directions. As indicated above this is a huge class of spaces: For instance any Polish (i.e. completely metrizable and separable) space is Borel (see [28, Theorem A1.2] and [11, Theorem A.47]) which is again a huge class even containing any topological space whose topology is locally compact, second-countable and Hausdorff (see e.g. [59, Sätze 8.15, 10.15, 13.17]). For u-measurable functions and sets we have powerful *projection* and *section* theorems at our disposal:

Theorem 2.1 (projections and sections). *Let S be a measurable space, T a Borel space and $A \in \mathcal{S} \otimes \mathcal{T}$.*

(i) *The projection of A on S, i.e. the set*

$$\mathrm{pr}_S A := \{s \in S : (s, t) \in A \text{ for some } t \in T\}$$

    is u-measurable.

(ii) *There is a u-measurable function $f : S \to T$ such that $(s, f(s)) \in A$ for all $s \in \mathrm{pr}_S(A)$.*

Proof. See [16, p. 252] or [18, p. 392]. $\square$

The next lemma highlights the advantages of u-measurability even further.

Lemma 2.2 (range and weak inverse). *Let S and T be Borel spaces and $f : S \to T$ be measurable.*

(i) *The image $f(S)$ of f is u-measurable;*

(ii) *There is a u-measurable function $g : T \to S$ satisfying $f \circ g \circ f = f$.*

Proof. The following proofs are taken from [31, Lemma 2.5].

(i) The graph of f is the set $\mathrm{Graph}(f) := \{(s, f(s)) : s \in S\}$. It is a measurable subset of $S \times T$ since it is the preimage of the (measurable) diagonal $\{(t, t) : t \in T\} \subset T^2$ under $\Phi : S \times T \to T^2$, $\Phi(s, t) = (f(s), t)$. Hence we may apply the Projection Theorem 2.1 (i) to the T-projection of $\mathrm{Graph}(f)$ and it remains to note that $\mathrm{pr}_T(\mathrm{Graph}(f)) = f(S)$.

(ii) Since T is Borel, the Section Theorem 2.1 (ii) yields a u-measurable function $g : T \to S$ such that $(g(t), t) \in \mathrm{Graph}(f), t \in f(S)$. Hence for $t \in f(S)$ put $s := g(t)$ and note that $t = f(s)$, which implies $f \circ g(t) = t$ for all such t - hence $f \circ g \circ f = f$. $\square$

2.1.2   Kernels and their regularity properties

A fundamental object for probability and measure theory is the following: If $(S, \mathcal{S})$ and $(T, \mathcal{T})$ are measurable spaces, a *kernel* from S to T is a map $\mu : S \times \mathcal{T} \to [0, \infty]$ with the properties

(1) $\mu(s, \cdot)$ is a measure on $(T, \mathcal{T})$ for any fixed $s \in S$,

(2) $\mu(\cdot, A)$ is a measurable map for any fixed $A \in \mathcal{T}$.

We sometimes write $\mu_s(\cdot)$ instead of $\mu(s, \cdot)$. Similarly a map $\mu : S \times \mathcal{T} \to [0, \infty]$ satisfying (1) and the modified property (2) where 'measurable' is replaced by 'universally measurable' is called *u-kernel* from S to T. A kernel μ is called *Markovian* or *stochastic* if $\mu_s(T) = 1, s \in S$, and *finite* if $\mu_s(T) < \infty, s \in S$. We call a kernel μ from S to T s-finite, if there is a sequence of finite kernels μ_n with $\mu_n \uparrow \mu$. A kernel μ is *pointwise σ-finite* if for each $s \in S$ the measure $\mu(s, \cdot)$ is a σ-finite measure on T and *uniformly σ-finite* if there is a partition $B_1, B_2, \ldots$ of T which simultaneously splits the μ_s into finite pieces, i.e. $\mu(s, B_i) < \infty, s \in S, i \in \mathbb{N}$. In between these two concepts, we call a kernel μ from S to T *σ-finite* if for each $s \in S$ there is a measurable partition $B_1^s, B_2^s, \ldots$ of T such that $(s, t) \mapsto \mathbf{1}\{t \in B_i^s\}$ is measurable for all $i \in \mathbb{N}$ and $\mu(s, B_i^s) < \infty, s \in S$. Clearly, a uniformly σ-finite kernel is σ-finite and any such kernel in turn is pointwise σ-finite. Just as there is a functional characterization of σ-finiteness of measures we have the following functional characterizations of the latter two types of kernels.

Lemma 2.3 (regularity properties of kernels). *Let μ be a kernel from S to T. The following holds:*

(i) *μ is σ-finite iff there exists a measurable map $f : S \times T \to (0, \infty)$ such that $\mu_s f(s, \cdot) < \infty, s \in S$ (in this case f may be chosen such that $\mu_s f(s, \cdot) < 1, s \in S$).*

(ii) *μ is pointwise σ-finite iff there exists a map $f : S \times T \to (0, \infty)$ such that the maps $f(s, \cdot), s \in S$, are measurable and $\mu_s f(s, \cdot) < \infty, s \in S$.*

Proof. (i) Let μ be a σ-finite kernel from S to T, i.e. for each $s \in S$ there is a partition $B_1^s, B_2^s, \ldots$ of T such that $(s, t) \mapsto \mathbf{1}\{t \in B_i^s\}$ is measurable for any $i \in \mathbb{N}$ and $\mu_s(B_i^s) < \infty$. Then we define

$$0 < f(s, t) := \sum_{i > 0} \frac{1}{1 + \mu(s, B_i^s)} \frac{1}{2^i} \mathbf{1}\{t \in B_i^s\} \leq 1$$

and note that

$$\mu_s f(s, \cdot) = \sum_{i > 0} \frac{\mu(s, B_i^s)}{1 + \mu(s, B_i^s)} \frac{1}{2^i} < 1.$$

Conversely, fix $f > 0$ on $S \times T$ with $\mu_s f(s, \cdot) < \infty$. Define

$$B_i^s := \left\{ t \in T : \frac{1}{i + 1} \leq f(s, t) < \frac{1}{i} \right\}, \quad i \in \mathbb{N},$$

and note that indeed since $0 < f < \infty$ this defines a partition of T for every $s \in S$. In addition

$$
\begin{aligned}
\mu(s, B_i^s) &= (i+1) \int \mathbf{1}\left\{\frac{1}{i+1} \leq f(s,t) < \frac{1}{i}\right\} \frac{1}{i+1} \mu(s, dt) \\
&\leq (i+1) \int \mathbf{1}\left\{\frac{1}{i+1} \leq f(s,t) < \frac{1}{i}\right\} f(s,t) \mu(s, dt) \\
&\leq (i+1) \mu_s f(s, \cdot) < \infty.
\end{aligned}
$$

(ii) Use the same construction as in part (a) (which boils down to proving the well-known criterion for σ-finite measures for every fixed $s \in S$). $\qquad\square$

2.2 Haar and invariant measures

In this section we summarize some basic facts concerning topological groups, Haar measure and invariance of measures and sets under group operations. We will also present some quickly accessible new results in Lemma 2.6, 2.8, 2.9 and 2.12.

2.2.1 Haar measure and modular function

By G we denote a group with neutral element e. Elements of a group G will usually be denoted by g or h. If G carries a σ-algebra $\mathcal{G}$ such that the maps $(g, h) \mapsto gh$ and $g \mapsto g^{-1}$ are measurable, then we call G a *measurable* group. Similarly, if G carries a topology $\mathcal{O}$ such that the above maps are continuous we call it a *topological* group. Clearly, any topological group becomes a measurable one when endowed with the Borel-σ-algebra $\mathcal{B}(\mathcal{O}) := \sigma(\mathcal{O})$. It is clear that left- or right-shifts on G are homeomorphisms in the topological and Borel isomorphisms in the measurable setting, such that translates $gA := \{ga : a \in A\}$ for $g \in G$ and $A \in \mathcal{G}$ are again measurable. A measure λ defined on $\mathcal{G}$ is *left-invariant* if

$$
\lambda(gA) = \lambda(A), \quad g \in G, A \in \mathcal{G},
$$

(*right-invariance* is defined with the obvious modification). In the topological setting, a classical result is that whenever the group carries a locally compact Hausdorff topology, then there is an up to constant multiples uniquely determined left-invariant *Radon* measure $\lambda \neq 0$ defined on its Borel σ-algebra. Here the Radon property means that λ is *locally finite* in the sense that it assigns finite values to compact sets, is *outer regular* on all Borel sets, meaning

$$
\lambda(A) = \inf\{\lambda(U) : U \supset A, U \text{ open}\}, \quad A \in \mathcal{B}(\mathcal{O}),
$$

and *inner regular* on all open sets, meaning

$$
\lambda(A) = \sup\{\lambda(K) : K \subset A, K \text{ compact}\}, \quad A \in \mathcal{O}.
$$

Such a measure λ is called a *(left) Haar measure* on the group G with respect to the topology $\mathcal{O}$. The following theorem constitutes a cornerstone of measure theory on groups and harmonic analysis. A proof may be found in [20, Theorems 11.8, 11.9].

Theorem 2.4 (existence and uniqueness of Haar measure). *Any locally compact Hausdorff group G possesses an up to positive multiples uniquely determined Haar measure.*

It is crucial to note at this point that the Radon condition enforces an intimate and important relation between topology and measure. As an example, consider the reals $\mathbb{R}$ on the one hand with the natural metric topology generated by $d(x, y) := |x - y|, x, y \in \mathbb{R}$, and on the other hand equipped with the discrete topology such that any subset $A \subset \mathbb{R}$ is open. Both topologies are locally compact and Hausdorff but the respective Haar measures differ and are given as (multiples of) Lebesgue measure in the first case and (multiples of) counting measure in the second.

The above example $\mathbb{R}$ with discrete topology and hence counting measure on its power set as corresponding Haar measure also shows that Haar measures need not be σ-finite in general. To enforce σ-finiteness of a Haar measure λ it is enough, for instance, to require the topology on G to be second-countable. Then, there is a countable partition of G into relatively compact subsets, on each of which λ is finite by definition.

If the topology on G is assumed to be second-countable in addition, then one may drop the inner and outer regularity conditions of Haar measure stated above - they will be implied by the local finiteness. In fact many authors consider Haar measure and Radon measures only in this more specific second-countable setting, e.g. [28, p. 36, p. 41].

In this thesis λ will always denote a σ-finite Haar measure on a group G, which in turn is always assumed to be locally compact, Hausdorff and second-countable, to ensure the existence of such a λ. We will abbreviate these conditions by saying that G is *lcsc*, the 'Hausdorff' condition being understood to be contained in the 'locally compact' condition. We now fix a left Haar measure λ on a lcsc group G. The left-invariance property may be rephrased using the usual tandem consisting of step functions and monotone convergence by writing

$$\int f(hg)\lambda(dg) = \int f(g)\lambda(dg), \quad h \in G, f \in \mathcal{G}_+.$$

For fixed $h \in G$ the measure $\lambda_h : A \mapsto \lambda(Ah)$ is by associativity of group multiplication again left-invariant, clearly non-zero and locally finite (note that left- and right-shifts on G are homeomorphisms). Hence, by uniqueness of Haar measure, there is a unique positive constant, which we call $\Delta(h)$, such that $\lambda_h = \Delta(h)\lambda$. The map $\Delta : G \to (0, \infty), h \mapsto \Delta(h)$ is called the *modular function* (though λ was needed in the construction of Δ here it is clear that Δ does not depend on the choice of normalization of λ). It even constitutes a continuous (in particular measurable) homomorphism $\Delta : G \to (0, \infty)$ (see e.g. [20, Prop. 11.10]) satisfying

$$\int f(gh)\lambda(dg) = \Delta(h^{-1}) \int f(g)\lambda(dg), \quad h \in G, f \in \mathcal{G}_+, \qquad (2.1)$$

and has the additional property that

$$\int f(g^{-1})\lambda(dg) = \int \Delta(g^{-1})f(g)\lambda(dg), \quad f \in \mathcal{G}_+. \qquad (2.2)$$

A group G is called *unimodular* if $\Delta(g) = 1$ for all $g \in G$. By (2.1) G is unimodular if and only if λ is right-invariant. Two examples of classes of unimodular groups are the Abelian groups and the compact groups (while the Abelian case is immediate note for compact G that the continuity and homomorphism property of Δ imply that the set $\Delta(G)$ must be a compact multiplicative subgroup of $(0, \infty)$ - the only such subgroup evidently being $\{1\}$).

Remark 2.5. For many results in this thesis topological requirements are subordinate and we might as well just assume G to be measurable and such that it carries a left-invariant σ-finite measure $\lambda \neq 0$. Kallenberg chooses this consequent and more natural setting in this recent paper [31] and in particular proves the existence of a measurable, topology-free variant of the modular function satisfying (2.1) and (2.2) ([31, Lemma 2.3]). As the existence of left-invariant σ-finite measures $\lambda \neq 0$ is still open without topological extra assumptions we shall always consider topological groups in this thesis.

2.2.2   Group operations and invariance

Let G be a group and S a set. An *operation* of G on S is given by a map $\varphi : G \times S \to S$, which we abbreviate by $gs := \varphi(g, s), g \in G, s \in S$, satisfying both $es = s, s \in S$, where e denotes the neutral element of G, and $g(hs) = (gh)s, g, h \in G, s \in S$. If such an operation along with the relevant map is given without risk of confusion then we omit the map and simply say that G *operates* or *acts* on S. In this case we also write $G \hookrightarrow S$.

Whenever the group G is measurable with σ-algebra $\mathcal{G}$ and the set S is a measurable space carrying a σ-algebra $\mathcal{S}$ we naturally require an operation of G on S to be *measurable*, in the sense that the underlying map $G \times S \to S$ is $\mathcal{G} \otimes \mathcal{S}/\mathcal{S}$-measurable. This condition implies the measurability of the *projections* $\pi_s : G \to S, s \in S$, and *shifts* $\theta_g : S \to S, g \in G$, given by

$$\pi_s(g) = \theta_g(s) = gs, \quad g \in G, s \in S.$$

Whenever G and S are topological spaces, we call an operation $G \hookrightarrow S$ *continuous* whenever the underlying map from $G \times S$ to S is. Clearly, continuous operations are measurable with respect to Borel σ-fields and the continuity passes on to projections and shifts.

For fixed $s \in S$ the set $\pi_s(G) = Gs$ is called the *orbit* of s and whenever $S = Gs$ for some $s \in S$ (and hence for all $s \in S$) the operation is called *transitive* and possesses only one orbit. Note that transitivity may be rephrased by saying that for any $s, t \in S$ there is always a $g \in G$ with $t = gs$.

We now consider a measurable operation $G \hookrightarrow S$ where G is lcsc with σ-finite Haar measure λ. It is interesting to note the following measurability property of the orbits, which holds in large generality.

Lemma 2.6 (u-measurability of orbits). *Given any measurable operation $G \hookrightarrow S$ where G is Borel, the orbits are u-measurable subsets of S.*

Proof. Consider the Borel isomorphism $\psi : G \times S \to G \times S$ given by $\psi(g, s) = (g, gs)$ and the measurable sets $A_s := \psi(G \times \{s\}), s \in S$. Since G is Borel the projection of

A_s on S is a u-measurable set in S according to Theorem 2.1 (i). These projections are clearly the orbits of the operation.   □

A subset $A \subset S$ is called *G-invariant* if $gA = A, g \in G$, and similarly a measure μ on S *G-invariant*, if

$$\mu(gA) = \mu(A), \quad g \in G, A \in \mathcal{S}.$$

If μ lives on a product space $S \times T$ with given operations $G \hookrightarrow S$ and $G \hookrightarrow T$, we call it *jointly G-invariant* if it is invariant with respect to the diagonal operation

$$g(s,t) := (gs, gt), \quad g \in G, s \in S, t \in T.$$

Furthermore, for $s \in S$,

$$G_{s,s} := \{g \in G : gs = s\} = \pi_s^{-1}(\{s\})$$

denotes the *stabilizer of s* and, taking a second $t \in S$,

$$G_{s,t} := \{g \in G : gs = t\} = \pi_s^{-1}(\{t\}),$$

is either empty (if $t \notin Gs$) or otherwise a left coset of $G_{s,s}$ since $G_{s,t} = g_{s,t}G_{s,s}$, where $g_{s,t}$ is a fixed group element satisfying $g_{s,t}s = t$. These subsets of G are measurable if the one point sets $\{t\}, t \in S$, are measurable. This is the case, for instance, if S is Borel.

2.2.3   Proper operations

The set of all σ-finite G-invariant measures on S is a convex cone in the sense that for any such μ, ν and $a, b \geq 0$ the measure $a\mu + b\nu$ is again σ-finite and G-invariant. The prime examples for G-invariant measures are the pushforwards $\lambda \circ \pi_s^{-1}, s \in S$, of the Haar measure λ under projections. But without further regularity assumptions on the operation, they need not be σ-finite. To enforce this, we assume that G operates *properly* on S in the sense that the operation is both measurable and such that the set of all pushforwards

$$\mu_s := \lambda \circ \pi_s^{-1}, \quad s \in S,$$

is even uniformly σ-finite. We recall that this requires the existence of a measurable partition $B_1, B_2, \ldots$ of S such that $\mu_s(B_n) < \infty, s \in S, n \in \mathbb{N}$. This concept was introduced by Kallenberg in [30] and generalizes the classical notion of a topologically proper operation (i.e. a continuous operation such that $\pi_s^{-1}(K) \subset G$ is compact whenever $K \subset S$ is) of a lcsc group on a lcsc space. He also showed (see [30, Lemma 2.1]) that properness is equivalent to the existence of a measurable function $k : S \to (0, \infty)$ such that

$$\mu_s k = \int k(t)\mu_s(dt) < \infty, \quad s \in S. \tag{2.3}$$

As indicated earlier topologically proper operations are proper in this more general sense.

Examples. (i) Trivially, compact groups operate properly on any measurable space if the operation is measurable. If they operate continuously on a topological space then the action is clearly topologically proper.

(ii) One readily proves that for $1 \leq u \leq d$ any u-dimensional linear subspace L of $\mathbb{R}^d$ operates on $\mathbb{R}^d$ via translation topologically proper.

(iii) $\mathbb{Z}^d$ operates on $\mathbb{R}^d$ via translation topologically proper.

(iv) $\mathbb{R}$ operates on the cylinder $Z := \mathbb{R} \times S^1$, S^1 denoting the 1-dimensional unit circle, via translation in the first component. This operation is trivially topologically proper.

(v) The *group of rigid motions* G_d operates topologically proper on $\mathbb{R}^d$ but acts not (even) properly on the *affine Grassmannian* $A(d, k)$, the space of all k-dimensional affine subspaces of $\mathbb{R}^d$. The problem here is that the stabilizer of a k-dimensional affine space E contains all translations with vectors contained in E, i.e. cannot be compact. As we will show in Corollary 3.10, proper operations with (even only locally) closed stabilizers automatically have compact stabilizers.

(vi) Given a graph $\Gamma = (V, E)$ with countable set of vertices V and edge set $E \subset V \times V$ its automorphism group $G = \mathrm{Aut}(\Gamma)$, endowed with a suitable topology, acts topologically proper on V in the canonical way

$$\mathrm{Aut}(\Gamma) \times V \to V, \quad (\varphi, s) \mapsto \varphi(s).$$

We postpone the details to Subsection 5.4.1.

(vii) Given a Riemannian manifold M, its isometry group $G = I(M)$, endowed with a suitable topology, acts topologically proper on M in the canonical way

$$I(M) \times M \to M, \quad (\varphi, s) \mapsto \varphi(s).$$

This is also true for closed subgroups of G (an example is (iv)). Again, details are postponed, in this case to Subsection 7.2.2.

2.2.4 Notions related to invariance of measures

The convex cone of all σ-finite (resp. s-finite) G-invariant measures on a Borel space S, where $G \hookrightarrow S$ is proper, has a striking structure that has been illuminated by Kallenberg in [30, Theorem 2.4] (resp. [31, Theorem 4.2]). We consider the σ-finite case first. As noted before the projection measures μ_s, $s \in S$, are invariant measures on S. They have the additional property that

$$\mu_{gs} = \Delta(g^{-1})\mu_s, \quad g \in G, s \in S, \tag{2.4}$$

which means that the properness condition upon the operation enforces

$$\Delta(g) = 1, \quad g \in G_{s,s}, s \in S. \tag{2.5}$$

Note that by (2.1) this means nothing but right $G_{s,s}$-invariance of λ. Choosing a system O of representatives of the orbits Gs, $s \in S$, the space S splits into the disjoint union

$$S = \bigcup_{b \in O} Gb,$$

and we may consider the *choice function* (in Kallenberg's related paper [31] called *orbit selector*) $\beta : S \to S$ where $\beta(s)$ denotes the previously fixed representative of Gs. In order to establish measurability of β, Gentner and Last [21] had to require measurability of O. The argument is the following:

Lemma 2.8 (u-measurability of selectors). *If $O \in \mathcal{S}$, then β is u-measurable.*

Proof. Note that for $B \subset S$ we have $\beta^{-1}(B) = G(B \cap O)$. Hence, recalling the Borel-Isomorphism $\varphi : G \times S \to G \times S$, $(g, s) \mapsto (g, gs)$, we have $\beta^{-1}(B) = G(B \cap O) = \mathrm{pr}_S(\varphi(G \times (B \cap O)))$ and for $B, O \in \mathcal{S}$ this implies together with Theorem 2.1 (i) that $\beta^{-1}(B)$ is u-measurable since G is Borel. $\qquad\square$

Kallenberg instead required S to be Borel and the operation to be proper (note that the above argument does not need any of these assumptions) and proved the existence of a u-measurable choice function without previously fixing a system of representatives of the orbits (which is then clearly given by the image of the choice function). The argument he gave in [31, Theorem 2.4] relies on Lemma 2.2 (ii), an invariant labeling of the orbits by a suitable kernel and the crucial Borel property of the space of probability measures on S which is inherited from S. It seems to be an open problem if the space of all σ-finite or even s-finite measures on a Borel space also inherits the Borel property.

Instead of using the function k from (2.3) some calculations simplify when k is replaced by the following normalized version w:

Lemma 2.9. (existence of a G-symmetric function) *Let G operate properly on S and assume that there is a measurable system of orbit representatives, such that we may fix one and call it O. Then there is a u-measurable function $w : S \to (0, \infty)$ on S such that*

$$\mu_b w = 1, \quad b \in O. \tag{2.6}$$

Proof. Since $G \hookrightarrow S$ is proper we may choose by [30, Lemma 2.1 (i)] a strictly positive measurable function $k : S \to (0, \infty)$ such that $\mu_b(k) < \infty, b \in O$. Now

$$w(s) := \frac{k(s)}{\mu_{\beta(s)}(k)}, \quad s \in S,$$

is u-measurable by Lemma 2.8 and Fubini's theorem and, in addition,

$$\mu_b w = \int \frac{k(s)}{\mu_b k} \mu_b(ds) = 1, \quad b \in O. \qquad\square$$

Note that if G is not unimodular it would not be consistent to require $\mu_s w = 1$ for all $s \in S$ by (2.4) which makes the restriction to O in (2.6) necessary. We now reformulate a result of Kallenberg [30, Theorem 2.4].

Theorem 2.10 (ergodic decomposition of invariant measures, Kallenberg).
Suppose G operates properly on the Borel space S and fix a measurable system O of orbit representatives. Then for any σ-finite, G-invariant measure ν on S there is a unique measure ν^* concentrated on O satisfying

$$\nu(\cdot) = \int \mu_b(\cdot)\nu^*(db). \tag{2.7}$$

Proof. Choosing k as in (2.3) the measures $\varphi_s := \mu_s/\mu_s k, s \in S$ are G-invariant, uniformly normalized in the sense that $\varphi_s k = 1$, $s \in S$, and even constant on orbits, i.e.

$$\varphi_{gs} = \varphi_s, \quad s \in S, g \in G. \tag{2.8}$$

In other words, the map $s \mapsto \varphi_s$ is a labeling of the orbits which is in addition measurable by Fubini's theorem, i.e. φ is a kernel on S. Kallenberg proved that this kernel can be used as a (normalized) extremal generator of the convex cone of all σ-finite invariant measures on S since any such measure ν on S may be written as (cf. [30, Theorem 2.4])

$$\nu(\cdot) = \int \varphi_s(\cdot)k(s)\nu(ds). \tag{2.9}$$

We now search a representation as in (2.9) which does not depend on the choice of k. As in Gentner and Last [21] we note, using (2.8) and (2.9), that

$$\nu(\cdot) = \int \varphi_s(\cdot)k(s)\nu(ds) = \int \mu_{\beta(s)}(\cdot)\mu_{\beta(s)}(k)^{-1}k(s)\nu(ds)$$
$$= \int \mu_b(\cdot)\nu^*(db), \tag{2.10}$$

where $\nu^* := (\mu_\beta(k)^{-1}k \cdot \nu) \circ \beta^{-1}$ is a σ-finite measure on S concentrated on O, in the sense that any measurable $B \subset S$ being disjoint with O has $\nu^*(B) = 0$. In spite of its definition ν^* is independent of k since ν^* is already uniquely determined by ν: To see this suppose ν_1^*, ν_2^* both satisfy (2.10) and are concentrated on O. Then

$$\iint f(s)\mu_b(ds)\nu_1^*(db) = \iint f(s)\mu_b(ds)\nu_2^*(db), \quad f \in \mathcal{S}_+,$$

and putting $f(s) := w(s)h(\beta(s))$ where w is as in (2.6) and $h \in \mathcal{S}_+$ is arbitrary yields $\nu_1^* h = \nu_2^* h, h \in \mathcal{S}_+$, and hence $\nu_1^* = \nu_2^*$. $\qquad\square$

Example 2.11 (countable S). Let G operate properly on the countable space S. Our Lemma 3.8 states that this is the case if and only if

$$0 < \lambda(G_{s,s}) < \infty, \quad s \in S.$$

Take a G-invariant measure ν on S and fix a (countable) system O of orbit representatives. Since by left-invariance of λ

$$\mu_b = \sum_{s \in Gb} \lambda(G_{b,s})\delta_s = \sum_{s \in Gb} \lambda(G_{b,b})\delta_s, \quad b \in O,$$

it follows from (2.7) that (with respect to O)

$$\nu^* = \sum_{b \in O} \frac{\nu(\{b\})}{\lambda(G_{b,b})} \delta_b. \tag{2.11}$$

Turning to the convex cone of s-finite G-invariant measures, Kallenberg proved in [31, Theorem 4.2] a version of the above theorem using the above mentioned existent u-measurable orbit selector and the associated inversion kernel (which we will construct in Subsection 3.1.1). A technical feature of his stream of arguments is that he needs not require the existence of a measurable system of orbit representatives and derives the existence of an orbit selector in a non-constructive way. From an application viewpoint, this can be both blessing and curse: an advantage is that the existence of a measurable system of orbit representatives O needs not be checked. On the downside, one is left without any information about the range $\beta(S)$, i.e. the induced system of representatives. Our approach has the advantage that it gives full flexibility to choose a particularly convenient one among the in applications usually plenty existing and easily accessible measurable ones and this will turn out to be useful in later applications. After fixing $O \in \mathcal{S}$ the associated choice function is then simply given by Lemma 2.8.

Given an operation $G \hookrightarrow S$ two classes of subsets of S will play a special role at several places in this thesis. The first class consists of the G-invariant sets $A \in \mathcal{S}$, where G-invariance means

$$gA = A, \quad g \in G.$$

These sets form a σ-algebra which we denote by $\mathcal{I}$. The second class is the collection of G-symmetric sets $B \in \mathcal{S}$, where G-symmetry refers to the property

$$0 < \mu_b B = \mu_c B < \infty, \quad b, c \in O.$$

The latter collection is not even closed with respect to $\cap$, $\cup$ or ^{c}. The use of the defining property of its members, namely that they consist of finite and non-zero pieces of each orbit will become apparent once we use them in the following chapters. Given a G-symmetric set $B \subset S$ we may define its *width* as

$$\delta(B) := \mu_b B \tag{2.12}$$

where $b \in O$ is fixed (and arbitrary). We note here a simple property of these objects.

Lemma 2.12 (invariant measures on symmetric intersections). *Given a G-invariant measure ν on S, a G-invariant set $A \subset S$ and a G-symmetric set $B \subset S$ the relation*

$$\nu(A \cap B) = \nu^*(A)\delta(B) \tag{2.13}$$

holds.

Proof. The decomposition (2.7) implies

$$\nu(A \cap B) = \iint \mathbf{1}_A(gb)\mathbf{1}_B(gb)\lambda(dg)\nu^*(db),$$

and since A is G-invariant this means

$$\nu(A \cap B) = \iint \mathbf{1}_A(b)\mathbf{1}_B(gb)\lambda(dg)\nu^*(db) = \delta(B) \int \mathbf{1}_A(b)\nu^*(db),$$

where we used the G-symmetry of B in the last step. $\qquad\square$

Remarks and Examples. (i) It is evident that any G-invariant set is a union of orbits and that any such union is G-invariant.

(ii) There are two important extreme situations: The first are transitive operations: If $G \hookrightarrow S$ is transitive then $\mathcal{I} = \{\emptyset, S\}$ and a set $B \in \mathcal{S}$ is G-symmetric if and only if $0 < \mu_c(B) < \infty$ for one $c \in S$ (and then in this case for all $c \in S$). In the case that $G \hookrightarrow G$ via left-translation this means nothing but $0 < \lambda(B) < \infty$. The other extreme case is when $G = \{e\} \hookrightarrow S$. This operation is *totally non-transitive* in the sense that each point is its own orbit. Here the roles of G-invariant and G-symmetric sets are reversed compared to the transitive case: the only G-symmetric set is S (the empty set is excluded for technical reasons) while any subset $A \subset S$ is now G-invariant.

It will be important for applications later to keep these extreme situations in mind.

(iii) $SO(d) \hookrightarrow \mathbb{R}^d$: Here $SO(d)$-invariant sets are given by all possible different unions of any of the concentric circles around the origin with arbitrary radii. Examples of G-symmetric sets are drafted in the following Figure 2.1.

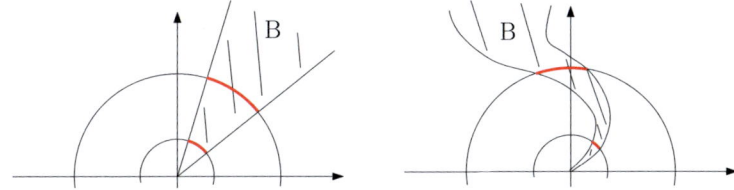

Figure 2.1: Examples of $SO(2)$-symmetric subsets B of $\mathbb{R}^2$.

(iv) $L \hookrightarrow \mathbb{R}^d$ where L is a fixed k-dimensional linear subspace of $\mathbb{R}^d$ where $k \in \{1, ..., d\}$: L-invariant sets are unions of parallel translates of L. Examples of L-symmetric sets are drafted in Figure 2.2.

(v) $\mathbb{Z}^d \hookrightarrow \mathbb{R}^d$: Any $\mathbb{Z}^d$-invariant set A may be represented by means of a uniquely determined subset $A_0 \subset [0, 1)^d$ such that

$$A = A_0 + \mathbb{Z}^d.$$

The collection of G-symmetric sets is given by the collection of all finite unions of the integer translates $z + [0, 1)^d, z \in \mathbb{Z}^d$. The width of such a set is just the number of translates the set consists of.

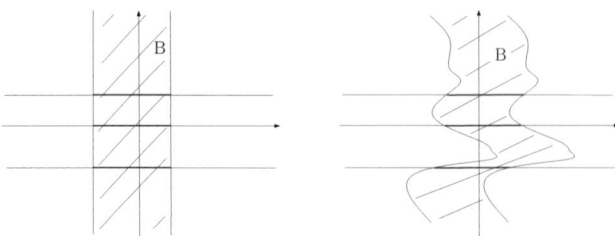

Figure 2.2: Examples of L-symmetric subsets B of $\mathbb{R}^2$ where $L = \{(x, 0) : x \in \mathbb{R}\}$.

2.3   Disintegration

Disintegrations emerge at many different places in Analysis, Probability Theory and especially in Stochastic Geometry. There are numerous examples, and among these are Cavalieri's principle, the calculus of conditional expectations and probabilities and the classical Palm formalism that leads to the notion of e.g. typical objects of stationary particle processes and many other meaningful objects. In this section we summarize the state of the art concerning existence of (invariant) disintegrations in the most general form known to the author. New is Lemma 2.15 and the elaborate existence result in Lemma 2.17 establishing measurably labeled invariant disintegrations of measurably labeled jointly invariant measures on product spaces.

2.3.1   Kernels and invariance

If G operates on both S and T and μ is a kernel from S to T then μ is G-invariant if

$$\mu(gs, gA) = \mu(s, A), \quad s \in S, A \in \mathcal{T}, g \in G. \tag{2.14}$$

The covariance property

$$\int f(t)\mu(gs, dt) = \int f(gt)\mu(s, dt), \quad g \in G, f \in \mathcal{T}_+,$$

is equivalent to (2.14) which is the reason why 'invariance' of kernels is sometimes referred to as 'covariance' or 'equivariance' in the literature. Given measures ν_1 and ν_2 on S and T respectively their product measure on $S \times T$ is denoted by $\nu_1 \otimes \nu_2$. Conversely, given a measure M on $S \times T$, M is usually not a product measure. Still, for a large class of such measures a similar decomposition, called *disintegration*, is possible - either from S to T in terms of one measure ν on S and a kernel μ from S to T or vice versa from T to S where the measure lives on T and the kernel is from T to S. A disintegration of M from S to T then reads, given ν and μ,

$$Mf = \iint f(s, t)\mu(s, dt)\nu(ds) =: (\nu \otimes \mu)f, \quad f \in (\mathcal{S} \otimes \mathcal{T})_+.$$

2.3.2   Disintegration on product spaces

In Probability Theory disintegrations arise for instance whenever a joint distribution of two random elements η and τ is conditioned on one variable. Then

$$\mathcal{L}(\eta, \tau) = \mathcal{L}(\eta) \otimes \mathbb{P}(\tau \in \cdot | \eta = \cdot) =: \mathcal{L}(\eta) \otimes \mathcal{L}(\tau | \eta = \cdot),$$

hence $\nu := \mathcal{L}(\eta)$ and $\mu(s, \cdot) := \mathbb{P}(\tau \in \cdot | \eta = s)$ is a valid disintegration as above with a probability measure ν and a Markovian kernel μ. Kallenberg [30] derived the existence of disintegrations for σ-finite measures M on product spaces using the procedure mentioned in Subsection 2.1.1:

Lemma 2.14. (disintegration of σ-finite measures) *A measure M on $S \times T$, where T is Borel is σ-finite iff there is a σ-finite measure ν on S and a σ-finite kernel μ from S to T such that*

$$M = \nu \otimes \mu.$$

Here μ may be chosen Markovian iff $M(\cdot \times T)$ is σ-finite.

Proof. See [30, Lemma 3.1] for the implication that a σ-finite M admits such a disintegration. The converse is most easily seen by invoking Lemma 2.3. The last assertion is trivial as we may choose $\nu = M(\cdot \times T)$ in this case. $\qquad \square$

We may even extend this result slightly to the s-finite case: note that any s-finite measure M on $S \times T$ where T is Borel may be disintegrated by means of a probability measure ν on S and an *s-finite* kernel μ from S to T. Here an s-finite kernel is a kernel that admits a sequence of finite kernels μ_n with $\mu_n \uparrow \mu$ or equivalently may be written as a countable sum of finite kernels.

Lemma 2.15 (disintegration of s-finite measures). *Let M be an s-finite measure on $S \times T$. Then $M = \nu \otimes \mu$ for a finite measure ν on S and an s-finite kernel μ from S to T. Here M is σ-finite iff μ may be chosen σ-finite. Further given a σ-finite measure $\tilde{\nu}$ on S such that $M(\cdot \times T) \ll \tilde{\nu}$ there is a suitable s-finite kernel $\tilde{\mu}$ from S to T with $M = \tilde{\nu} \otimes \tilde{\mu}$. As above M and $\tilde{\mu}$ are simultaneously σ-finite.*

Proof. The case $M = 0$ is trivial such that we may assume $M \neq 0$. If $M = \sum_{n \geq 1} M_n$ with finite non-zero measures M_n on $S \times T$ we have disintegrations $M_n = \nu_n \otimes \mu_n$ with finite non-zero measures ν_n and finite kernels μ_n by Lemma 2.14 (or simply, after obvious modifications, the existence of conditional distributions). Now define the probability measure

$$\nu := \sum_{n \geq 1} \frac{1}{2^n \nu_n(S)} \nu_n.$$

Since $\nu_n \ll \nu$, $n \in \mathbb{N}$, and each measure under consideration is finite, there are Radon-Nikodym densities $f_n : S \to [0, \infty)$ with $\nu_n = f_n \cdot \nu$. Using these, we may define the s-finite kernel μ from S to T via

$$\mu(s, \cdot) := \sum_{n \geq 1} f_n(s) \mu_n(s, \cdot).$$

Then the monotone convergence theorem yields

$$\nu \otimes \mu = \sum_{n \geq 1} \nu \otimes f_n \mu_n = \sum_{n \geq 1} (f_n \cdot \nu) \otimes \mu_n = \sum_{n \geq 1} \nu_n \otimes \mu_n = \sum_{n \geq 1} M_n = M.$$

If M is σ-finite, then Lemma 2.14 yields a disintegration $M = \nu \otimes \mu$ with σ-finite ν and σ-finite μ. Using a function $f > 0$ on S with $\nu f < \infty$ we may rewrite this disintegration via $M = \nu \otimes \mu = (f \cdot \nu) \otimes (\frac{1}{f} \mu)$ which yields the desired disintegration. The converse implications are trivial.

For the last assertion, fix $\tilde{\mu}$ with the stated properties and consider a fixed disintegration $M = \nu \otimes \mu$. Here ν and μ may be chosen such that $\mu(s,T) > 0, s \in S$, since $A := \{s \in S : \mu(s,T) > 0\}$ is measurable and we may form $\mathbf{1}_A \cdot \nu$ and redefine μ outside of A suitably. Then clearly $\nu \sim M(\cdot \times T) \ll \tilde{\nu}$ and the Radon-Nikodym Theorem yields a measurable function $f \geq 0$ on S with $\nu = f \cdot \tilde{\nu}$. Putting $\tilde{\mu}(s,\cdot) := f(s)\mu(s,\cdot), s \in S$, yields

$$\tilde{\nu} \otimes \tilde{\mu} = \tilde{\nu} \otimes (f\mu) = (f \cdot \tilde{\nu}) \otimes \mu = \nu \otimes \mu = M. \qquad \square$$

2.3.3   Invariant disintegrations

Note that if G operates on both S and T and the disintegration $M = \nu \otimes \mu$ consists of a G-invariant measure ν on S and a G-invariant kernel μ from S to T, then M is *jointly G-invariant*, in the sense that M is invariant with respect to the diagonal operation $g(s,t) := (gs, gt), g \in G, s \in S, t \in T$, of G on $S \times T$, i.e.

$$\int f(gs, gt) M(d(s,t)) = Mf, \quad g \in G, f \in (\mathcal{S} \otimes \mathcal{T})_+.$$

Conversely, it is natural to ask if a σ-finite, jointly G-invariant measure M on $S \times T$ admits such an *invariant disintegration* where both ν and μ are G-invariant. This is in fact a problem that has been in the focus of many authors since the 1960's. Two main approaches were successful in different contexts: the *skew factorization* approach of Matthes [45] (1963) and the combined *regularization* and *perfection* approach which appeared first in the paper [61] by Ryll-Nardzewski in 1961. The classical skew factorization of a jointly invariant measure on a product space requires one of the two factors, say S, to be the group G. Then the bijective skew-shift $\vartheta(g,t) := (g, gt)$ transforms the jointly invariant measure M on $G \times T$ into the measure $M \circ \vartheta$ on $G \times T$ which is invariant with respect to shifts in the first component (only). It is then only a small step to deduce that any such measure is a product measure of the form $\lambda \otimes \rho$ with a σ-finite measure ρ on T, and reversing the skew shift then gives the desired disintegration. Kallenberg significantly generalized this approach in [31] by showing how this technique may be even applied in the general setting for jointly invariant measures on $S \times T$ by using the inversion kernel. We shall give a short summary of his ideas in Subsection 3.2.1 for two reasons: First these nicely support the relevance of the inversion kernel which is part of this thesis and second we shall need them to establish Theorem 3.9 (which seems new in this generality and will be needed in this form later).

On the other hand the regularization and perfection approach is more elaborate: one first needs to identify an invariant *supporting measure* ν on S, i.e. a σ-finite invariant measure satisfying $M(\cdot \times T) \ll \nu$. This is difficult since it may happen that $M(\cdot \times T)$ is not σ-finite. Then a complicated construction follows (see our Lemma 2.17 which contains this as a special case) which comprehends the regularization of a family of Radon-Nikodym densities as well as an averaging procedure over G smoothing the resulting kernel into an invariant one. Gentner and Last used this technique to construct the inversion kernel in [21] and we shall present this construction in Subsection 3.1.1.

In his 2007 paper [30, Section 3] Kallenberg extended and compared both methods and gave some applications to Palm (and related) kernels. Using the regularization and perfection approach he proves in [30] the following theorem:

Theorem 2.16. (invariant disintegration of σ-finite invariant measures, Kallenberg)

(i) *A σ-finite measure M on $S \times T$, where T is Borel, is jointly G-invariant if and only if it admits an invariant disintegration from S to T.*

(ii) *If M as in (i) is jointly G-invariant and $\nu \gg M(\cdot \times T)$ is σ-finite and G-invariant then there is a σ-finite and G-invariant kernel μ from S to T with $M = \nu \otimes \mu$.*

(iii) *If in (ii) $M(\cdot \times T)$ is σ-finite, then we may choose $\nu := M(\cdot \times T)$ and the associated G-invariant μ is stochastic.*

Proof. (i) A proof of the implication that any jointly G-invariant M admits such a disintegration may be found in [30, Corollary 3.6]. The converse follows from a simple calculation. (ii) is [30, Theorem 3.5] and (iii) is trivial. $\square$

We will also give a complete proof of an extension of this result to measurably labeled families of jointly invariant measures on product spaces in Lemma 2.17. In addition we will further extend this theorem to the case of s-finite M in Theorem 3.9.

2.3.4 Invariant disintegration of kernels

Given measurable spaces R, S and T and operations $G \hookrightarrow S$ and $G \hookrightarrow T$, our aim in this subsection is to prove a measurable and invariant decomposition of measurably labeled families of jointly G-invariant measures $\{M_r\}_{r \in R}$ on $S \times T$ in an invariant and measurable way. For this we need the following lemma which is a crucial extension of known results on the existence of disintegrations of measures on product spaces (see e.g. [28, Theorem 6.3]) and their respective G-invariant versions for jointly G-invariant measures found by Kallenberg in [30]. Though the proof is a straightforward adaption of arguments found in [28, p. 107] and the regularization and perfection arguments in [30, Theorem 3.5] it will serve as our main tool in the construction of the inversion kernel in Subsection 3.1.1.

Lemma 2.17 (invariant disintegrations of kernels). *Let R, S, T be measurable spaces where S and T are Borel, M a σ-finite kernel from R to $S \times T$ and let G operate measurably on both S and T.*

(i) *There is a stochastic kernel ν from R to S and a σ-finite kernel κ from $R \times S$ to T such that, writing $\kappa_r := \kappa(r, \cdot, \cdot)$,*

$$M_r = \nu_r \otimes \kappa_r, \quad r \in R.$$

(ii) *If ν' is a σ-finite kernel from R to S with $M_r(\cdot \times T) \ll \nu'_r, r \in R$, then there is a σ-finite kernel κ' from $R \times S$ to T such that*

$$M_r = \nu'_r \otimes \kappa'_r, \quad r \in R.$$

(iii) *If M is such that M_r is jointly G-invariant for each $r \in R$ and ν is a σ-finite kernel from R to S such that each $\nu_r, r \in R$, is a G-invariant measure on S with $M_r(\cdot \times T) \ll \nu_r$ for $r \in R$, then there is a σ-finite kernel κ from $R \times S$ to T with the invariance property, writing $\kappa_r := \kappa(r, \cdot, \cdot)$,*

$$\kappa_r(gs, A) = \kappa_r(s, \theta_g^{-1}A), \quad A \in \mathcal{T}, s \in S, g \in G, r \in R,$$

such that

$$M_r = \nu_r \otimes \kappa_r, \quad r \in R.$$

Proof. (i) We may assume that $M_r(S \times T) > 0, r \in R$. Since M is σ-finite we may choose by Lemma 2.3 a measurable function $f > 0$ on $R \times S \times T$ such that $M_r f_r = 1$, $r \in R$, and define the stochastic kernel P from R to $S \times T$ as $P_r := f_r \cdot M_r$, $r \in R$. Then [28, Proposition 7.26] yields a stochastic kernel $\tilde{\kappa}$ from $R \times S$ to T such that together with the stochastic kernel $\nu_r := P_r(\cdot \times T)$

$$P_r = \nu_r \otimes \tilde{\kappa}_r, \quad r \in R,$$

c.f. Dellacherie/Meyer [16, 5.58]. This is clearly equivalent to

$$M_r = \nu_r \otimes \kappa_r, \quad r \in R,$$

where $\kappa(r, s, A) := \int \mathbf{1}_A(t) f(r, s, t)^{-1} \tilde{\kappa}(r, s, dt)$, $A \in \mathcal{T}$, and thus proves the first assertion.

(ii) If ν' is a given kernel from R to S with the property $M_r(\cdot \times T) \ll \nu'_r, r \in R$, then ν_r from above satisfies $\nu_r \sim M_r(\cdot \times T) \ll \nu'_r$, $r \in R$, and by Dellacherie/Meyer [16, 5.58] we may choose a measurable function $f : R \times S \to [0, \infty]$ such that

$$f(r, s) = \frac{d\nu_r}{d\nu'_r}(s), \quad \nu'_r\text{-a.e. } s \in S.$$

Then

$$M_r = \nu'_r \otimes \kappa'_r, \quad r \in R,$$

where $\kappa'(r, s, \cdot) := f(r, s)\kappa(r, s, \cdot)$, which proves the second statement.

(iii) From (ii) we get a kernel κ from $R \times S$ to T with $M_r = \nu_r \otimes \kappa_r, r \in R$. Invariance of M_r and ν_r imply for any $f \in (\mathcal{S} \otimes \mathcal{T})_+$ that

$$\iint f(s, t)\kappa_{r,gs}(dt)\nu_r(ds) = \iint f(s, t)\kappa_{r,s} \circ \theta_g^{-1}(dt)\nu_r(ds), \quad g \in G, r \in R.$$

Since T is Borel it admits in particular a countable measure determining class which gives

$$\kappa_{r,gs} = \kappa_{r,s} \circ \theta_g^{-1}, \quad \nu_r\text{-a.e. } s \in S, g \in G, r \in R.$$

Fixing some right Haar measure $\tilde{\lambda}$ on G Fubini's theorem yields in particular

$$\kappa_{r,gs} = \kappa_{r,s} \circ \theta_g^{-1}, \quad \tilde{\lambda}\text{-a.e. } g \in G, \nu_r\text{-a.e. } s \in S, r \in R. \tag{2.15}$$

Let $l > 0$ be some measurable function on G with $\tilde{\lambda}l = 1$. Then we may define

$$\bar{\kappa}_{r,s} := \int (\kappa_{r,hs} \circ \theta_h)(l \cdot \tilde{\lambda})(dh). \tag{2.16}$$

A similar calculation as in [30, Theorem 3.5] shows that on the sets

$$A_r := \{s \in S : \kappa_{r,ps} \circ \theta_p = \kappa_{r,qs} \circ \theta_q, \; \tilde{\lambda}^2\text{-a.e.}(p,q) \in G^2\}, \quad r \in R,$$

we have

$$\overline{\kappa}_{r,s} = \overline{\kappa}_{r,hs} \circ \theta_h, \quad h \in G, s \in A_r, r \in R. \tag{2.17}$$

We now show that the map $(r,s) \mapsto \mathbf{1}_{A_r}(s)$ is measurable. For this take a countable measure determining class $\mathcal{C}$ of $\mathcal{T}$ and define for each $B \in \mathcal{C}$ the measurable map

$$m_B(r,p,q) := \min\{\kappa_{r,ps} \circ \theta_p(B), \kappa_{r,qs} \circ \theta_q(B)\}, \quad r \in R, p,q \in G.$$

Then also

$$z_B(r,s) := \iint l(p)l(q)|(\kappa_{r,ps} \circ \theta_p - \kappa_{r,qs} \circ \theta_q)(B)|\mathbf{1}\{m_B(r,p,q) < \infty\}\tilde{\lambda}^2(d(p,q))$$

is measurable by Fubini's Theorem. Further we clearly have $s \in A_r$ iff for each $B \in \mathcal{C}$ the terms $z_B(r,s)$ are zero. Hence we may write

$$\mathbf{1}_{A_r}(s) = 1 - \mathbf{1}\{\sup_{B \in \mathcal{C}} z_B(r,s) > 0\},$$

which reveals the desired measurability. Further one can easily check that A_r is G-invariant and (2.15) implies that $\nu_r(A_r^c) = 0$. Finally we define

$$\tilde{\kappa}_{r,s} := \mathbf{1}_{A_r}(s)\overline{\kappa}_{r,s}, \quad s \in S, r \in R. \tag{2.18}$$

Then by invariance of A_r and (2.17)

$$\tilde{\kappa}_{r,gs}(A) = \tilde{\kappa}_{r,s}(g^{-1}A), \quad g \in G, s \in S, A \in \mathcal{T}, r \in R,$$

and since $\tilde{\kappa}_{r,s} = \overline{\kappa}_{r,s} = \kappa_{r,s}, \nu_r$-a.e. $s \in S$, the required disintegrations

$$M_r = \nu_r \otimes \tilde{\kappa}_r, \quad r \in R,$$

hold indeed. $\qquad\qquad\qquad\qquad\qquad\qquad\qquad\qquad\qquad\qquad\qquad\qquad\square$

Remark 2.18. Here the smoothing of the kernel κ in (2.16) is referred to as *regularization* while the selection of 'nice' kernel members in (2.18) is called *perfection* of the kernel.

2.4 Random measures and stationarity

In Subsection 2.4.1 we first define random measures in a very general (non-topological) framework following Kallenberg in [31] and present some of their properties. Then, in Subsection 2.4.2 we define and discuss group stationarity of random elements.

2.4.1   Random measures and Palm pairs

Let $(S, \mathcal{S})$ be a measurable space and $\mathbf{M}(S)$ the space of all σ-finite measures on S. We endow $\mathbf{M}(S)$ with the smallest σ-field $\mathcal{M}(S)$ rendering the mappings $\mu \mapsto \mu(B)$ for all $B \in \mathcal{S}$ measurable. Let $(\Omega, \mathcal{A}, \mathbb{P})$ be a σ-finite measure space. We use probabilistic notations even though $\mathbb{P}$ need not be a probability measure. In particular, we denote by $\mathbb{E}$ integration with respect to $\mathbb{P}$. A *random measure* on S is a measurable mapping $\xi : \Omega \to \mathbf{M}(S)$ that is σ-finite in the sense that for each $\omega \in \Omega$ there is a countable partition $B_1^\omega, B_2^\omega, \ldots$ of S such that $\xi(\omega, B_i^\omega) < \infty$ $\mathbb{P}$-a.e. $\omega \in \Omega$ for any $i \in \mathbb{N}$ and such that $(\omega, s) \mapsto \mathbf{1}\{s \in B_i^\omega\}$ is measurable for $i \in \mathbb{N}$, i.e. ξ is nothing but a σ-finite kernel from Ω to S using the kernel notation $\xi(\omega, B) := \xi(\omega)(B)$. A *point process* on S is a random measure on S which charges all measurable sets with values in $\mathbb{N} \cup \{0, \infty\}$. We denote the identity map on Ω by θ_e in order to be consistent with (2.26).

If ξ is a random measure on S then the *Campbell measure* C_ξ of ξ with respect to $\mathbb{P}$ is the measure on $\Omega \times S$ satisfying

$$C_\xi f = \mathbb{E} \int f(\theta_e, s) \xi(ds), \quad f \in (\mathcal{A} \otimes \mathcal{S})_+. \tag{2.19}$$

Further if η is a random element in a Borel space T then

$$C_{\xi, \eta} f = \mathbb{E} \int f(\eta, s) \xi(ds), \quad f \in (\mathcal{T} \otimes \mathcal{S})_+, \tag{2.20}$$

is called the *Campbell measure of the pair* (ξ, η). These measures have the following properties.

Lemma 2.19 (properties of Campbell measures). (i) *Given a random measure ξ on a measurable space S its Campbell measure C_ξ is σ-finite.*

(ii) *Given random measures ξ and $\tilde{\xi}$ on a Borel space S, then*

$$\xi = \tilde{\xi} \quad \mathbb{P}\text{-a.e.} \quad \Leftrightarrow \quad C_\xi = C_{\tilde{\xi}}.$$

(iii) *Given in addition a random element η in a Borel space T the Campbell measure of the pair (ξ, η) is s-finite and it is σ-finite whenever ξ is η-measurable or $\mathbb{E}\xi$ is σ-finite.*

Proof. (i) is evident in view of Lemma 2.3.

(ii) One implication of the equivalence is trivial. To see the other, suppose $C_\xi = C_{\tilde{\xi}}$, i.e.

$$C_\xi f = C_{\tilde{\xi}} f, \quad f \in (\mathcal{A} \otimes \mathcal{S})_+.$$

Since S is Borel there is a countable measure determining class $\mathcal{C} \subset \mathcal{S}$. For $B \in \mathcal{C}$ the special choice $f(\omega, s) = g(\omega)\mathbf{1}_B(s)$ for an arbitrary $g \in \mathcal{A}_+$ yields $\xi(B) = \tilde{\xi}(B)$ $\mathbb{P}$-a.e., and since $\mathcal{C}$ is countable this yields

$$\xi(B) = \tilde{\xi}(B), \quad B \in \mathcal{C}, \mathbb{P}\text{-a.e.}.$$

As $\mathcal{C}$ is measure determining this yields the assertion.

(iii) The s-finiteness is proved in [30, Lemma 4.2] as well as the σ-finiteness in the case when ξ is η-measurable. The case when $\mathbb{E}\xi$ is σ-finite is immediate as we may chose a function $f > 0$ on S such that $\mathbb{E}\xi f < \infty$ such that also $\int f(s) C_{\xi, \eta}(d(t, s)) = \mathbb{E}\xi f < \infty$. $\qquad \square$

The σ-finiteness of C_ξ does not necessarily carry over to the *intensity measure* $\mathbb{E}\xi$ of ξ defined via $(\mathbb{E}\xi)(A) := \mathbb{E}\xi(A), A \in \mathcal{S}$. Note that $\mathbb{E}\xi = C_\xi(\Omega \times \cdot)$ and that σ-finiteness is usually not preserved under projections. But $\mathbb{E}\xi$ is s-finite, since the class of s-finite measures *is* closed under projections - a key observation of Kallenberg in [30]. This makes sometimes the use of *supporting measures of* ξ necessary. These are σ-finite measures equivalent to $\mathbb{E}\xi$ in the sense of mutual absolute continuity.

Lemma 2.20 (s-finiteness and its use, Kallenberg).

  (i) *Any s-finite measure on a product space $S \times T$ has s-finite projections $M(\cdot \times T)$ and $M(S \times \cdot)$.*

 (ii) *Given a random measure ξ its intensity measure $\mathbb{E}\xi$ is s-finite.*

(iii) *For any s-finite measure ν there is a finite measure $\tilde{\nu} \sim \nu$.*

(iv) *Any random measure ξ possesses a finite supporting measure.*

Proof. (i) If $M_n \uparrow M$ is an approximating sequence of finite measures then $M_n(\cdot \times T)$ and $M_n(S \times \cdot)$ are approximating sequences of finite measures for $M(\cdot \times T)$ and $M(S \times \cdot)$, respectively. Now (ii) follows since C_ξ is σ-finite, in particular s-finite, and $\mathbb{E}\xi = C_\xi(\Omega \times \cdot)$. For (iii) we take a sequence of finite non-zero measures $\nu_n \uparrow \nu$ and note that $\tilde{\nu} := \sum_n 2^{-n} \nu_n(S)^{-1} \nu_n$ has the desired property. Now (iv) follows from (ii) and (iii). $\qquad\square$

Note that the definition of $\mathbb{E}\xi$ together with the monotone convergence theorem yields

$$\mathbb{E} \int f(s)\xi(ds) = \int f(s)\mathbb{E}\xi(ds), \quad f \in \mathcal{S}_+.$$

This identity will be used frequently and is called *Campbell's Theorem* in some parts of the literature. If Ω is Borel then Lemma 2.14 yields a σ-finite measure ν on S and a σ-finite kernel Q from S to Ω disintegrating C_ξ as follows:

$$C_\xi f = \iint f(\omega, s)Q_s(d\omega)\nu(ds), \quad f \in (\mathcal{A} \otimes \mathcal{S})_+. \tag{2.21}$$

We call any pair (ν, Q) satisfying (2.21) a *Palm pair* of ξ (see [21]). The kernel Q is the ν-associated *Palm kernel* of ξ. To make the dependence on ξ explicit, we sometimes write $(\nu_\xi, Q_\xi) := (\nu, Q)$. Q may be chosen to be stochastic if and only if $\mathbb{E}\xi$ is σ-finite in which case the measures Q_s are probability measures on Ω, the *Palm probability measures* on Ω. Since structural requirements on Ω are not desirable one may consider instead a random element η in a Borel space T and form similar to (2.21) by means of Lemma 2.15 (together with the fact that $C_{\xi,\eta}$ is always s-finite according to Lemma 2.19 (iii)) a disintegration of the form

$$C_{\xi,\eta} f = \iint f(t, s)P_s(dt)\nu(ds), \quad f \in (\mathcal{A} \otimes \mathcal{S})_+. \tag{2.22}$$

Given a disintegration as in (2.22) we call the P-kernel members ν-*associated Palm (pseudo) distributions* $P_s, s \in S$, of η. Whenever a ν-associated Palm kernel Q exists then clearly

$$P_s = Q_s(\eta \in \cdot) \quad \nu\text{-a.e.}$$

It is clear from the construction that P_s contains less information on the underlying stochastic experiment than Q_s which makes it sometimes preferable to work directly on Ω instead on T, see Subsection 2.4.2. If $\mathbb{E}\xi$ is σ-finite such that Q_s and P_s are $\mathbb{E}\xi$-a.e. stochastic then P_s is also called the *Palm distribution of η with respect to ξ at $s \in S$* and we shall write, following Kallenberg [32, 33], in this case

$$\mathbb{P}(\eta \in \cdot||\xi)_s := P_s(\cdot), \quad s \in S.$$

Of particular importance is the case when $\eta = \xi$. If there is a fixed partition of S into measurable sets $P := \{B_1, B_2, \dots\}$ such that $\xi(B_i) < \infty$ $\mathbb{P}$-a.e. then $\xi \in \mathbf{M}^P(S)$ $\mathbb{P}$-a.s. where

$$\mathbf{M}^P(S) := \{\mu \in \mathbf{M}(S) : \mu(B_i) < \infty, i \in \mathbb{N}\} \qquad (2.23)$$

is a measurable subset of $\mathbf{M}(S)$. It can be shown that $\mathbf{M}^P(S)$ is Borel whenever S is by following similar arguments as in [28, pp. 561, 564]. It seems to be an open problem whether or not the corresponding statement for $\mathbf{M}(S)$ is true. Again, if $\mathbb{E}\xi$ is σ-finite and ξ takes $\mathbb{P}$-a.s. values in a Borel space, then we may choose $\nu = \mathbb{E}\xi$ and (2.22) reads

$$\mathbb{E}\int f(\xi, s)\xi(ds) = \iint f(\mu, s)\mathbb{P}(\xi \in d\mu||\xi)_s(\mathbb{E}\xi)(ds), \quad f \in (\mathcal{M}(S) \otimes \mathcal{S})_+. \quad (2.24)$$

In contrast to our work [21] we will rarely use Palm pairs in this thesis (and only sometimes the Palm (pseudo)-distributions). Their existence is only insured if Ω is Borel and we shall not impose any such regularity conditions on our underlying abstract probability space. The better object to look at will be a certain measure $\mathbb{Q}$ on $\Omega \times O$ that we will introduce in Section 4.1.1 and whose existence does not depend on structural properties of Ω.

Replacing in (2.22) ξ by its n-fold random product measure $\xi^n = \xi \otimes \cdots \otimes \xi$ yields, under the assumption that $\mathbb{E}\xi^n$ is σ-finite, the *nth-order Palm distributions* $\mathbb{P}(\eta \in \cdot||\xi^n)_s$ *of η with respect to ξ*. In addition, if ξ is a point process on the Borel space S we may write $\xi = \sum_i \delta_{\tau_i}$ where the sum is taken either over a finite set of the form $\{1, \dots, n\}$ for some $n \in \mathbb{N}$, or over $\mathbb{N}$ itself with a corresponding (finite or infinite) sequence of random elements τ_i in S. In either case we may form the random measures

$$\xi^{(n)} := \sum_{(i_1, \dots, i_n)_{\neq}} \delta_{(\tau_{i_1}, \dots, \tau_{i_n})}, \quad n \in \mathbb{N}$$

on S^n respectively, where the sum is taken over all $(i_1, \dots, i_n) \in \mathbb{N}^n$ with pairwise different components. Their intensity measures are called the *factorial moment measures* of ξ (of order n respectively). We finally mention a technical lemma here.

Lemma 2.21 (transformations of random measures). *Let ξ denote a random measure on S, $f : \Omega \times S \to [0, \infty)$ a measurable function and γ a σ-finite kernel from $\Omega \times S$ to T. Then*

(i) *$\eta(\omega, \cdot) := \int \mathbf{1}\{s \in \cdot\}f(\omega, s)\xi(\omega, ds)$ is a random measure on S,*

(ii) $(\xi \otimes \gamma)(\omega, \cdot) := \iint \mathbf{1}\{(s,t) \in \cdot\}\gamma(\omega, s, dt)\xi(\omega, ds)$ *is a random measure on* $S \times T$.

(iii) ξ^n *is a random measure on* S^n.

(iv) *If* S *is Borel and* ξ *is a point process, then* $\xi^{(n)}$ *is a point process on* S^n.

Proof. (i) We need to show that η is a σ-finite kernel from Ω to S. Choose $h : \Omega \times S \to (0, \infty)$ such that $\xi(\omega, h(\omega, \cdot)) < \infty$ and let $A := \{(\omega, s) : f(\omega, s) > 0\}$. Then put

$$g(\omega, s) := \mathbf{1}_{A^c}(\omega, s) + \mathbf{1}_A(\omega, s)\frac{h(\omega, s)}{f(\omega, s)} > 0, \quad \omega \in \Omega, s \in S,$$

and observe that $\eta g(\omega) = \int \mathbf{1}\{(\omega, s) \in A\}h(\omega, s)\xi(\omega, ds) < \infty$. The assertion now follows from Lemma 2.3 (i).

(ii) Again we need to check measurable σ-finiteness. Choose $f_\delta : \Omega \times S \times T \to (0, \infty)$ with $\delta(s, s, f_\delta(\omega, s, \cdot)) < 1, \omega \in \Omega, s \in S$, and $f_\xi : \Omega \times S \to (0, \infty)$ with $\xi(\omega, f_\xi(\omega, \cdot)) < 1, \omega \in \Omega$. Putting $f := f_\delta f_\xi$ the assertion follows again from Lemma 2.3 (i).

For (iii) choose by Lemma 2.3 (i) a measurable function $f : \Omega \times S \to (0, \infty)$ such that $\xi(\omega, f(\omega, \cdot)) < 1$ and define $g : \Omega \times S^n \to (0, \infty)$ by

$$g(\omega, s_1, \ldots, s_n) := f(\omega, s_1) \ldots f(\omega, s_n)$$

then clearly $\xi^n(\omega, g(\omega, \cdot)) < 1$. The assertion now follows from Lemma 2.3 (i).

(iv) now follows from (iii) since $\xi^{(n)}(\omega, f(\omega, \cdot)) \leq \xi^n(\omega, f(\omega, \cdot))$ together with the fact that $\xi^{(n)}$ takes values in $\mathbb{N} \cup \{0, \infty\}$. $\qquad\square$

The next lemma is due to Mecke [46].

Lemma 2.22 (taming of random measures). *Let* ξ *be a random measure on the measurable space* S.

(i) There is a function $h : \Omega \times S \to (0, \infty)$ such that

$$\int h(\omega, s)\xi(\omega, ds) = \mathbf{1}\{\xi(\omega) \neq 0\}, \quad \omega \in \Omega.$$

(ii) If ξ is uniformly σ-finite with respect to a partition P, there is a function $h : \mathbf{M}^P(S) \times S \to (0, \infty)$ such that

$$\int h(\xi(\omega), s)\xi(\omega, ds) = \mathbf{1}\{\xi(\omega) \neq 0\}, \quad \omega \in \Omega.$$

Proof. (i) Choose by means of Lemma 2.3 a measurable $f : \Omega \times S \to (0, \infty)$ such that $\xi(\omega, f(\omega, \cdot)) < \infty$. Then

$$h(\omega, s) := \mathbf{1}\{\xi(\omega) \neq 0\}\frac{f(\omega, s)}{\int f(\omega, t)\xi(\omega, dt)} + \mathbf{1}\{\xi(\omega) = 0\}, \quad \omega \in \Omega, s \in S,$$

has the desired property.

(ii) If $P = \{B_1, B_2, \dots\}$ then we may put

$$a(\mu, s) = \sum_i 2^{-i} \frac{1}{1 + \mu(B_i)} \mathbf{1}_{B_i}(s), \quad \mu \in \mathbf{M}^P(S), s \in S,$$

and note that if $\mu \neq 0$ we have

$$0 < \int a(\mu, s)\mu(ds) < 1.$$

We may then define

$$h(\mu, s) := \mathbf{1}\{\mu \neq 0\} \frac{a(\mu, s)}{\int a(\mu, t)\mu(dt)} + \mathbf{1}\{\mu = 0\}, \quad \mu \in \mathbf{M}^P(S), s \in S,$$

which has the desired properties. $\square$

The last lemma in this subsection will be needed for our discussion of Cox-Delauney mosaics in Chapter 7.

Lemma 2.23 (conditioning with respect to an integrating measure). *Let ξ denote a random element in a Borel space T and let η denote a uniformly σ-finite random measure η on a measurable space S. Then for any $n \in \mathbb{N}$*

$$\mathbb{E} \int f(\xi, s)\eta^n(ds) = \mathbb{E} \iint f(t, s)\mathbb{P}(\xi \in dt | \eta)\eta^n(ds), \quad f \in (\mathcal{T} \otimes \mathcal{S}^n)_+,$$

where $\mathbb{P}[\xi \in \cdot | \eta]$ denotes a regular version of the conditional distribution of ξ given η.

Proof. As the uniform σ-finiteness of η clearly carries over to the random measure η^n, say with respect to the partition P of S^n, we choose by means of Lemma 2.22 (ii) a function $h : \mathbf{M}^P(S^n) \times S^n \to (0, \infty)$ with

$$\int h(\eta^n(\omega), s)\eta^n(\omega, ds) = \mathbf{1}\{\eta^n(\omega) \neq 0\} = \mathbf{1}\{\eta(\omega) \neq 0\}, \quad \omega \in \Omega.$$

We first show that for any $f \in (\mathcal{T} \otimes \mathcal{S}^n)_+$

$$\mathbb{E} \int f(\xi, s)h(\eta^n, s)\eta^n(ds) = \mathbb{E} \iint f(t, s)\mathbb{P}(\xi \in dt | \eta)h(\eta^n, s)\eta^n(ds). \qquad (2.25)$$

Both sides are measures in f and by monotone convergence, it is enough to show that

$$\mathbb{E} \int \mathbf{1}\{(\xi, s) \in A\}h(\eta^n, s)\eta^n(ds) = \mathbb{E} \iint \mathbf{1}\{(t, s) \in A\}\mathbb{P}(\xi \in dt | \eta)h(\eta^n, s)\eta^n(ds),$$

for all $A \in \mathcal{T} \otimes \mathcal{S}^n$. As both sides are finite (!) measures in A, we may, by a monotone class argument (or a uniqueness result such as [28, Lemma 1.17] which is essentially the same), further reduce to the case when $A = B \times C$ where $B \in \mathcal{T}$ and $C \in \mathcal{S}^n$. Starting left, we obtain by conditioning on η that

$$
\mathbb{E} \int \mathbf{1}\{\xi \in B, s \in C\} h(\eta^n, s) \eta^n(ds)
$$
$$
= \iint \mathbf{1}\{t \in B\} \int \mathbf{1}\{s \in C\} h(\mu^n, s) \mu^n(ds) \mathbb{P}(\xi \in dt | \eta = \mu) \mathbb{P}(\eta \in d\mu)
$$
$$
= \int \mathbb{P}(\xi \in B | \eta = \mu) \int \mathbf{1}\{s \in C\} h(\mu^n, s) \mu^n(ds) \mathbb{P}(\eta \in d\mu)
$$
$$
= \mathbb{E} \left[\mathbb{P}\left[\xi \in B | \eta \right] \int \mathbf{1}\{s \in C\} h(\eta^n, s) \eta^n(ds) \right].
$$

We may proceed via

$$
\mathbb{E} \int \mathbf{1}\{\xi \in B, s \in C\} h(\eta^n, s) \eta^n(ds) = \mathbb{E} \int \mathbf{1}\{s \in C\} \mathbb{P}(\xi \in B | \eta) h(\eta^n, s) \eta^n(ds)
$$
$$
= \mathbb{E} \int \mathbb{P}\left[\xi \in B, s \in C | \eta \right] h(\eta^n, s) \eta^n(ds),
$$

and thus we proved (2.25). Applying (2.25) to the random variable $\tilde{\xi} := (\xi, \eta)$ in the Borel space $T \times \mathbf{M}^P(S)$ yields for any measurable $f : T \times \mathbf{M}^P(S) \times S^n \to [0, \infty)$

$$
\mathbb{E} \int f(\xi, \eta, s) h(\eta^n, s) \eta^n(ds) = \mathbb{E} \iint f(t, \mu, s) \mathbb{P}((\xi, \eta) \in d(t, \mu) | \eta) h(\eta^n, s) \eta^n(ds).
$$

Choosing here finally

$$
f(t, \mu, s) = \frac{\tilde{f}(t, s)}{h(\mu^n, s)}, \quad t \in T, \mu \in \mathbf{M}^P(S), s \in S^n,
$$

for arbitrary $\tilde{f} \in (\mathcal{T} \otimes \mathcal{S})_+$ yields

$$
\mathbb{E} \int \tilde{f}(\xi, s) \eta^n(ds) = \mathbb{E} \iint \frac{\tilde{f}(t, s)}{h(\mu^n, s)} \mathbb{P}((\xi, \eta) \in d(t, \mu) | \eta) h(\eta^n, s) \eta^n(ds)
$$
$$
= \mathbb{E} \iint \tilde{f}(t, s) \mathbb{P}((\xi, \eta) \in d(t, \mu) | \eta) \eta^n(ds)
$$
$$
= \mathbb{E} \iint \tilde{f}(t, s) \mathbb{P}(\xi \in dt | \eta) \eta^n(ds),
$$

which is the assertion. $\qquad\square$

2.4.2 The canonical framework for stationarity

(Partial) Stationarity of a random measure refers to invariance of its distribution with respect to an operating group. To capture this description precisely, consider an operation $G \hookrightarrow S$ of some lcsc group G on some measurable space S. This operation induces an action $G \hookrightarrow \mathbf{M}(S)$ of G on the space of all σ-finite measures on S via

$$
g\mu(\cdot) := \mu \circ \theta_g^{-1}(\cdot) = \mu(g^{-1}\cdot), \quad g \in G, \mu \in \mathbf{M}(S).
$$

The reason for defining the shift of a measure in terms of a shift by g^{-1} rather than a shift by g is that this choice leads to the covariance property

$$
\int f(s)(g\mu)(ds) = \int f(gs)\mu(ds), \quad g \in G, f \in \mathcal{S}_+,
$$

by using monotone convergence. Now, a G-*stationary* random measure ξ is a random measure whose law $\mathcal{L}(\xi)$ is a G-invariant probability measure on $\mathbf{M}(S)$ with respect to the induced operation $G \hookrightarrow \mathbf{M}(S)$, i.e.

$$\mathbb{P}(\xi \in gA) = \mathbb{P}(\xi \in A), \quad g \in G, A \in \mathcal{M}(S).$$

Evidently this is equivalent to saying that $g\xi$ has the same distribution as ξ for each $g \in G$. Stationarity is a purely distributional property and is independent of the concrete functional representation of ξ as a random element of $\mathbf{M}(S)$. Still, among the many possible representations of ξ as a map from a space Ω to $\mathbf{M}(S)$ there is one that is particularly useful and convenient for Palm calculus: Choosing the canonical setting $\Omega := \mathbf{M}(S)$ the identity map $\xi(\omega) := \omega$ becomes a random measure with distribution $\mathbb{P}$. Hence, G-stationarity of ξ is nothing but G-invariance of $\mathbb{P}$ in this setting. Evidently the relation

$$\xi(g\omega) = g\omega = g\xi(\omega)$$

holds in addition. Hence stationarity may (without any loss of generality) be represented by the following mathematical framework.

Assume that G operates measurably on Ω (we do not need to require any further regularity conditions here and in fact e.g. assuming properness here leads to heavy inconsistencies: the operation $G \hookrightarrow \mathbf{M}(S)$ is far from proper for non-compact G) and write $\theta_g\omega := g\omega$. The reason for this sudden change of notation is that it enables us to distinguish between group elements themselves and actual shifts on Ω, which become Ω-valued random variables in this setting. The family $\{\theta_g : g \in G\}$ is referred to as *flow* on Ω in the literature, and the induced structural properties of this flow

$$\theta_e\omega = \omega, \quad \omega \in \Omega, \quad \text{and} \quad \theta_g \circ \theta_h = \theta_{gh}, \quad g, h \in G, \tag{2.26}$$

are often called *flow-properties*. The canonical setting mentioned above motivates the following assumption and definition: We assume that $\mathbb{P}$ is invariant under the flow and a G-stationary random measure on S is a random element $\xi : \Omega \to \mathbf{M}(S)$ satisfying

$$\xi(\theta_g\omega) = g\xi(\omega), \quad g \in G, \omega \in \Omega,$$

which means set wise

$$\xi(\theta_g\omega, B) = g\xi(\omega, B) = \xi(\omega, g^{-1}B), \quad g \in G, \omega \in \Omega, B \in \mathcal{S}. \tag{2.27}$$

Note that a G-stationary random measure is in this frame nothing but a G-invariant kernel, which is the reason why some authors call stationary random measures *invariant random measures*.

Some additional words seem adequate to highlight the advantages of the above framework. In Palm Theory the underlying stochastic experiment is usually a G-stationary quite complicated random element, e.g. a stationary random set, tessellation, particle process or random measure. The random measure ξ with respect to which the Palm measure is then formed is a derivate of this 'underlying stochastic experiment' which is *entirely* captured by ω but *not* by $\xi(\omega)$ since information might be lost. For instance ω might live in the space of tessellations of $\mathbb{R}^d$, while $\xi(\omega)$ is the point process of the vertices of this tessellation. In probability theory only the distribution of a random element is of interest, usually not the functional representation in terms of a specific space Ω. Choosing the canonical setting means to restrict oneself to such a functional representation but the shifts on Ω itself have the huge advantage, that they happen at the level where *all* the information about the stochastic experiment is available. This simplifies formalities when looking at two or more derived 'typical' objects of a stationary process.

Similarly to stationarity of random measures we may define stationarity of random elements. If G operates measurably on a space T and τ is a random element in T, then τ is *G-stationary* if $\mathcal{L}(\tau)$ is a G-invariant measure. By similar arguments as above we may choose Ω such that τ satisfies

$$\tau(\theta_g \omega) = g\tau(\omega), \quad \omega \in \Omega, g \in G,$$

and $\mathbb{P}$ to be G-invariant. *Joint G-invariance* of several random elements $\tau_1, ..., \tau_n$ is defined as invariance of their joint distribution with respect to the diagonal operation of G on the product of the spaces, where these random elements live. Again, we may choose Ω such that

$$(\tau_1(\theta_g \omega), \ldots, \tau_n(\theta_g \omega)) = g(\tau_1(\omega), \ldots, \tau_n(\omega)), \quad \omega \in \Omega, g \in G,$$

and choose $\mathbb{P}$ to be G-invariant.

Chapter 3

Inversion kernel and applications

This chapter is devoted to the construction of an important kernel in Section 3.1 and to further constructions and conclusions based on this kernel in Section 3.2.

3.1 Inversion kernel

We will give a construction of the *inversion kernel* in Subsection 3.1.1 which is taken from our paper [21] and thus constitutes an original part of this thesis. This kernel first appeared in a paper by Rother and Zähle [60] established in the setting of homogeneous spaces, i.e. transitive operations of topological groups on topological spaces with some topological regularity assumptions. In [31] Kallenberg (independently of Gentner and Last) established the existence of the inversion kernel for possibly non-transitive group actions at about the same time as well. Kallenberg's approach for constructing this kernel is very different from ours and the interested reader might wish to read more about his elegant construction in [31, Theorem 3.1].

We shall then give examples of operations and respective inversion kernels in Subsection 3.1.2.

3.1.1 Construction of the inversion kernel

In the following Theorem 3.1 we introduce a kernel κ from $S \times S$ to G that will enable us to handle stabilizers and their cosets within G in integrals with respect to Haar measure λ on G. This kernel satisfies

$$\int f(gs,g)\lambda(dg) = \iint f(t,g)\kappa_{s,t}(dg)\mu_s(dt), \quad f \in (\mathcal{S} \otimes \mathcal{G})_+, s \in S. \tag{3.1}$$

In particular κ disintegrates the Haar measure λ on G along each orbit via

$$\int f(g)\lambda(dg) = \iint f(g)\kappa_{s,t}(dg)\mu_s(dt), \quad f \in \mathcal{G}_+, s \in S. \tag{3.2}$$

Theorem 3.1 (inversion kernel). *If G operates properly on the Borel space S there is a unique kernel κ from $S \times S$ to G satisfying (3.1) and with the properties*

(i) $\kappa_{s,gt} = \kappa_{s,t} \circ \theta_g^{-1}, \quad g \in G, s, t \in S,$

(ii) $\kappa_{s,t}$ *is concentrated on $G_{s,t} := \{g \in G : gs = t\}$ for $t \in Gs, s \in S$,*

(iii) $\kappa_{s,t}(G) = 1, \quad t \in Gs, s \in S.$

Proof. Consider the kernel

$$M_s := \int \mathbf{1}\{(gs, g) \in \cdot\}\lambda(dg), \quad s \in S,$$

from S to $S \times G$ which is clearly measurably σ-finite by the properness of $G \hookrightarrow S$ and has the property that every M_s is a jointly G-invariant measure on $S \times G$. Further it is clear that $\mu_s = \lambda \circ \pi_s^{-1} = M_s(\cdot \times G)$ and since the μ_s are σ-finite G-invariant measures we may apply Lemma 2.17 with $R := S, T := G$ and $\nu_s := \mu_s$ to the kernel M to obtain a kernel κ from $S \times S$ to G such that (3.1) and the invariance property (i) are fulfilled. It remains to show that κ fulfills (ii) and (iii): For (ii) note that for $s \in S$ by (3.1) (and the Borel property of S which insures measurability of the relevant indicators)

$$\iint \mathbf{1}\{gs \neq t\}\kappa_{s,t}(dg)\mu_s(dt) = \int \mathbf{1}\{gs \neq gs\}\lambda(dg) = 0.$$

This means that

$$\kappa_{s,t}(G_{s,t}^c) = 0, \quad \mu_s\text{-a.e. } t \in S, s \in S,$$

and since $\mu_s \neq 0$ for each $s \in S$ we may pick some $t \in Gs$ such that $\kappa_{s,t}(G_{s,t}^c) = 0$ holds. But then by (i) $\kappa_{s,t}(G_{s,t}^c) = 0$ for all $t \in Gs$ (if $\tilde{t} \in Gs$ then $\tilde{t} = ht$ for some $h \in G$). For (iii) choose k as in (2.3) and note that setting $f(t, g) := k(t)$ in (3.1) yields

$$\mu_s k = \int k(t)\kappa_{s,t}(G)\mu_s(dt) = \kappa_{s,s}(G)\mu_s k, \quad s \in S, \tag{3.3}$$

where we applied (i) in the last step. Again by (i) this implies $\kappa_{s,t}(G) = 1$ for $t \in Gs$.

To prove uniqueness of κ suppose there is another kernel $\tilde{\kappa}$ with the desired properties. Then in particular

$$\iint f(t, g)\kappa_{s,t}(dg)\mu_s(dt) = \iint f(t, g)\tilde{\kappa}_{s,t}(dg)\mu_s(dt), \quad f \in (\mathcal{S} \otimes \mathcal{G})_+, s \in S.$$

Since G is Borel (in particular $\mathcal{G}$ is countably generated) this implies

$$\kappa_{s,t} = \tilde{\kappa}_{s,t}, \quad \mu_s\text{-a.e. } t \in S, s \in S,$$

and the invariance property (i) of both κ and $\tilde{\kappa}$ yields $\kappa_{s,t} = \tilde{\kappa}_{s,t}, t \in Gs, s \in S$. Finally, by (ii) we may conclude $\kappa = \tilde{\kappa}$ since $G_{s,t} = \emptyset$ for all $t \notin Gs$. $\square$

In this thesis we will use the inversion kernel in the following form exclusively.

Corollary 3.2 (one-parametric version). *Let $G \hookrightarrow S$ be proper and assume that a measurable system of orbit representatives exists. Fixing such a system O with associated u-measurable choice function β the map*

$$(s, B) \mapsto \kappa_{\beta(s),s}(B), \quad s \in S, B \in \mathcal{G},$$

is a u-kernel from S to G.

Proof. Just note that the map $(s, t) \mapsto \kappa_{s,t}(B)$ is measurable according to Theorem 3.1 while $s \mapsto (\beta(s), s)$ is u-measurable by Lemma 2.8 and elementary properties of the product σ-algebra. $\square$

3.1.2   Special actions and their inversion kernels

We will investigate special cases and examples of proper operations along with their respective inversion kernels in this section.

Example 3.3 (transitive case). A first step of specialization is to assume that the operation $G \hookrightarrow S$ is transitive. Here $c := \beta(s)$, $s \in S$, is just one single representative and since $G_{c,s} \neq \emptyset$ for all $s \in S$ the corresponding measure $\kappa_s := \kappa_{c,s}$ is never 0. If in addition the stabilizer $G_{c,c}$ is locally compact (which is inherited from G for instance if $G_{c,c}$ is *locally closed*, i.e. the intersection of an open and a closed subset of G, see also [10, I.65]), then κ_c is nothing but Haar measure with total mass 1 on $G_{c,c}$ and κ_s is a translate of this measure representing the from $G_{c,c}$ uniformly distributed mass shifted onto the coset $G_{c,s}$. Note that this implies that necessarily a proper transitive operation which is topologically well-behaving in the sense that stabilizers are at least locally closed, must have compact stabilizers as these carry a finite Haar measure. We refer to the following Corollary 3.10 for a detailed proof of a generalization of this statement in the general non-transitive setting.

Example 3.4 (group case). Assume the lcsc G operates on itself via left-translation. This is clearly a transitive operation which is even continuous. It is clearly topologically proper as for compact $K \subset G$ the sets $\pi_s^{-1}(K) = Ks^{-1}$ are trivially again compact. Also the sets $G_{e,s} = \{s\}$, $s \in G$, are compact. Here we have $\mu_s = \Delta(s^{-1})\lambda$, $s \in G$. Further we may choose $O = \{e\}$, $\beta(g) = e$, $g \in G$, and since $G_{e,s} = \{s\}$, $s \in G$, we have

$$\kappa_s = \kappa_{e,s} = \delta_s, \quad s \in S.$$

Example 3.5 (trivial operation). Consider the trivial operation $G = \{e\} \hookrightarrow S$ where S is an arbitrary measurable space. This operation is clearly proper since $\{e\}$ is compact and we have $O = S$ (there is only this choice) $\lambda = \delta_e$, $\beta(s) = s$, and $\mu_s = \delta_s$, $s \in S$. Further $G_{\beta(s),s} = G_{s,s} = \{e\} = G$ and the inversion kernel reduces to

$$\kappa_{\beta(s),s} = \kappa_{s,s} = \delta_e, \quad s \in S.$$

It is clear that any measure or kernel is invariant with respect to this operation.

Example 3.6 (countable G). Consider the measurable operation $G \hookrightarrow S$ where G is a countable group endowed with the discrete topology and S an arbitrary measurable space. Here the Haar measures are all constant multiples of counting measure (which is clearly also right-invariant, hence any such G is unimodular). Choosing λ as counting measure on G, we get

$$\mu_s = \int \mathbf{1}\{gs \in \cdot\}\lambda(dg) = \sum_{g \in G}\delta_{gs} = \sum_{t \in Gs}|G_{s,t}|\delta_t = |G_{s,s}|\sum_{t \in Gs}\delta_t, \quad s \in S.$$

Here we used in the last step left-invariance of counting. As we may clearly find a function $k > 0$ on S such that $\sum_{t \in Gs}k(t) < \infty$, $s \in S$, this means that $G \hookrightarrow S$ is proper iff $|G_{s,s}| < \infty$, $s \in S$ (note that $0 < |G_{s,s}|$, $s \in S$, since $e \in G_{s,s}$). In this setting evidently the inversion kernel is given by

$$\kappa_{\beta(s),s} = \frac{1}{|G_{\beta(s),\beta(s)}|}\sum_{g \in G_{\beta(s),s}}\delta_g, \quad s \in S.$$

Note that the mass of $\kappa_{\beta(s),s}$ on a point $g \in G$ is either $1/|G_{\beta(s),s}| = 1/|G_{\beta(s),\beta(s)}|$ if $g \in G_{\beta(s),s}$ and 0 otherwise. When s varies within a fixed orbit these point masses wander from one coset to the next, but the masses themselves do not change. But they may change when s jumps from one orbit to another since the number of elements in the cosets may vary.

Example 3.7 (countable S). Consider the measurable operation $G \hookrightarrow S$ where G is lcsc and S is a countable space with the power set as respective σ-algebra. Here the following lemma characterizes properness. As before, the cardinality of a set A is denoted by $|A|$.

Lemma 3.8 (proper operations on countable sets). *G operates properly on a countable set S if and only if*

$$0 < \lambda(G_{s,s}) < \infty, \quad s \in S.$$

In this case

$$\frac{\lambda(G_{s,s})}{\lambda(G_{\beta(s),\beta(s)})} = \frac{|G_{s,s}\beta(s)|}{|G_{\beta(s),\beta(s)}s|}, \quad s \in S, \tag{3.4}$$

and either all orbits are infinite or all orbits are finite.

Proof. The countability of S implies $0 < \lambda(G_{s,s})$, $s \in S$, since $\lambda(G_{s,s}) = 0$ for some s enforces

$$\lambda(G) = \sum_{t \in Gs} \lambda(G_{s,t}) = \sum_{t \in Gs} \lambda(G_{s,s}) = 0$$

by left-invariance of λ which is impossible. For any $k : S \to [0, \infty)$ on S we have by left-invariance of λ

$$\mu_s k = \int k(gs) \sum_{t \in Gs} \mathbf{1}\{gs = t\} \lambda(dg) = \lambda(G_{s,s}) \sum_{t \in Gs} k(t), \quad s \in S. \tag{3.5}$$

Hence if $G \hookrightarrow S$ is proper then choosing $k > 0$ as in (2.3) shows that $\lambda(G_{s,s}) < \infty$ for any $s \in S$. Equation (3.5) also shows the converse since we may always choose $k > 0$ on S such that $\sum_{t \in Gs} k(t) < \infty$, $s \in S$.

Now assume that $G \hookrightarrow S$ is proper. By left-invariance of Haar measure we have

$$\lambda(G_{s,s}) = \lambda(G_{s,s} \cap G_{t,t})|G_{s,s}t|,$$

which implies, since $0 < \lambda(G_{s,s}) < \infty$ that $0 < |G_{s,s}t|, \lambda(G_{s,s} \cap G_{t,t}) < \infty, s, t \in S$. Putting $t = \beta(s)$ and dividing the resulting equality with the same equality where s and $\beta(s)$ are interchanged yields (3.4). To see the last assertion note that for any orbit Gs we have $\lambda(G) = |Gs|\lambda(G_{s,s})$ and hence if $|Gt| = \infty$ for some $t \in S$, then necessarily $\lambda(G) = \infty$ and thus for any other orbit Gs also $|Gs| = \infty$ by properness. $\qquad\square$

3.2 Consequences and applications

In order to illustrate the usefulness of the inversion kernel we will present some consequences for proper operations from its existence and some applications in this section. For our purposes Kallenberg's idea in [31] to combine the skew factorization approach with the inversion kernel in order to derive invariant disintegrations of jointly invariant measures will be very useful to extend Theorem 2.16 to the s-finite case in Theorem 3.9. Besides the pure sake of generality there is the following reason to do this: intensity measures of random measures are not necessarily σ-finite, but they must be s-finite according to Lemma 2.20. Having established Theorem 2.16 in this generality will allow us to drop technical extra assumptions such as σ-finiteness of certain intensity measures in later theorems. In particular, this extension will be helpful for proving the stochastic version of the mass-transport principle in Theorem 5.5 in a shorter and more transparent way as we did in [21].

3.2.1 Disintegration revisited

We have derived the existence of the inversion kernel using Lemma 2.17. This lemma clearly contains the existence of invariant disintegrations of jointly invariant measures on product spaces: it is enough to specialize R to be a one point set. In [31] Kallenberg also makes heavy use of this result in his existence proof of the inversion kernel though his argumentation does not use something similar to our Lemma 2.17. Instead of purely considering (3.1), he integrated (3.1) against an arbitrary probability measure ν, and derived by the ordinary existence of invariant disintegrations a ν-associated inversion kernel γ^ν. Then he proves in an interesting second step that γ^ν is essentially independent of ν, see [31, Theorem 3.1]. Even though both constructions in [21] and [31] of the inversion kernel use the existence of invariant disintegration of jointly invariant measures, it is interesting to inspect invariant disintegrations again using the inversion kernel as this sheds some additional light on them.

What follows is a quick summary of Kallenberg's ideas in [31] where he combined the skew-factorization technique with the inversion kernel to derive invariant disintegrations:

Consider a jointly invariant measure M on a product space $S \times T$ where both factors are Borel. Then

$$\hat{M}f := \iint f(g, \beta(s), t)\kappa_{\beta(s),s}(dg)M(d(s,t)) \tag{3.6}$$

defines a measure on $G \times S \times T$, which is concentrated on $G \times O \times T$. This transformation is a bijection between jointly invariant measures on $S \times T$ and measures on $G \times O \times T$ that are invariant with respect to joint shifts in G and T only. Now using the bijective skew-shift

$$\vartheta(g, s, t) := (g, s, gt)$$

we may consider the measure $\hat{M} \circ \vartheta$ on the same space, which is now invariant with respect to shifts in G (only). Hence, the well-known factorization for such measures (see e.g. [31, Lemma 2.2]) yields a measure ρ on $S \times T$ (more precisely on $O \times T$)

such that $\hat{M} \circ \vartheta = \lambda \otimes \rho$. Now the point is that an arbitrary (!) disintegration $\rho = \hat{\nu} \otimes \hat{\mu}$ with a measure $\hat{\nu}$ on O and a kernel $\hat{\mu}$ from O to T yields an invariant disintegration $M = \nu \otimes \mu$ where

$$\nu(\cdot) := \iint \mathbf{1}\{gb \in \cdot\}\lambda(dg)\hat{\nu}(db) \tag{3.7}$$

and

$$\mu(s, \cdot) := \iint \mathbf{1}\{gt \in \cdot\}\hat{\mu}_{\beta(s)}(dt)\kappa_{\beta(s),s}(dg). \tag{3.8}$$

This may be verified by direct calculation: First note that ν is a G-invariant measure, μ is a G-invariant kernel (as follows from Theorem 3.1 (i)) and that

$$\iint f(s, t)\mu(s, dt)\nu(ds) = \iiint f(gb, t)\mu(gb, dt)\lambda(dg)\hat{\nu}(db)$$
$$= \iiint f(gb, gt)\mu(b, dt)\lambda(dg)\hat{\nu}(db)$$
$$= \iiiint f(gb, ght)\hat{\mu}(b, dt)\kappa_{b,b}(dh)\lambda(dg)\hat{\nu}(db)$$
$$= \iiint f(gb, gt)\hat{\mu}(b, dt)\lambda(dg)\hat{\nu}(db),$$

where we used Fubini and right $G_{b,b}$-invariance of λ (note that $\Delta(h^{-1}) = 1$ for $h \in G_{b,b}$) in the last step. Since $\lambda \otimes \hat{\nu} \otimes \hat{\mu} = \hat{M} \circ \vartheta$ we may proceed

$$\iint f(s, t)\mu(s, dt)\nu(ds) = \int f(gb, gt)\hat{M} \circ \vartheta(d(g, b, t)) = \int f(gb, t)\hat{M}(d(g, b, t)),$$

and by definition of $\hat{M}$ we arrive at

$$\iint f(s, t)\mu(s, dt)\nu(ds) = \iint f(g\beta(s), t)\kappa_{\beta(s),s}(dg)M(d(s, t)) = Mf.$$

Kallenberg's complete result in [31] says that the correspondence $M \leftrightarrow \hat{M}$ establishes a bijection between jointly invariant σ-finite measures M on $S \times T$ and σ-finite measures $\hat{M}$ on $G \times O \times T$ invariant with respect to joint shifts in the first and last component. Using these insights together with Lemma 2.15 we may establish the existence of invariant disintegrations even for jointly G-invariant s-finite measures:

Theorem 3.9 (invariant disintegrations of s-finite measures). *Let M be an s-finite jointly G-invariant measure on $S \times T$. Then there is a σ-finite G-invariant measure ν on S and an s-finite G-invariant kernel μ from S to T with $M = \nu \otimes \mu$. In addition, given a fixed G-invariant σ-finite measure $\tilde{\nu}$ on S such that $M(\cdot \times T) \ll \tilde{\nu}$ there is a suitable G-invariant s-finite kernel $\tilde{\mu}$ from S to T with $M = \tilde{\nu} \otimes \tilde{\mu}$. M and μ, resp. $\tilde{\mu}$, are simultaneously σ-finite. If $M(\cdot \times T)$ is σ-finite, then the $\nu := M(\cdot \times T)$-associated μ is stochastic.*

Proof. Define the s-finite measure $\hat{M}$ on $G \times O \times T$ by (3.6). The measure $\hat{M} \circ \vartheta$ is G-invariant with respect to shifts in G (only) and Lemma 2.2 in [31] yields $\hat{M} \circ \vartheta = \lambda \otimes \rho$ with an s-finite measure ρ on $O \times T$. Take a disintegration $\rho = \hat{\nu} \otimes \hat{\mu}$ by means of Lemma 2.15 with a finite measure $\hat{\nu}$ on S and an s-finite kernel $\hat{\mu}$ from O to T.

Define the σ-finite ν as in (3.7) and the s-finite μ as in (3.8). Exactly the same calculation as above shows that $M = \nu \otimes \mu$. For the last assertion, fix $\tilde{\mu}$ with the stated properties and consider a fixed invariant disintegration $M = \nu \otimes \mu$. Here ν and μ may be chosen such that $\mu(s, T) > 0, s \in S$, since $A := \{s \in S : \mu(s, T) > 0\}$ is G-invariant, and $\mathbf{1}_A \cdot \nu$ is thus again G-invariant. Then clearly $\nu \sim M(\cdot \times T) \ll \tilde{\nu}$ and [30, Lemma 2.3] yields a measurable G-invariant function $f \geq 0$ on S with $\nu = f \cdot \tilde{\nu}$. Putting $\tilde{\mu}(s, \cdot) := f(s)\mu(s, \cdot), s \in S$, yields

$$\tilde{\nu} \otimes \tilde{\mu} = \tilde{\nu} \otimes (f\mu) = (f \cdot \tilde{\nu}) \otimes \mu = \nu \otimes \mu = M.$$

The rest is evident in view of Lemma 2.15. $\qquad \square$

3.2.2 Disproving properness

In order to be able to apply parts of the theory in this thesis one needs to check properness for the concrete operation $G \hookrightarrow S$ which is of interest. If it is indeed proper then it is usually not hard to determine a suitable partition that splits the μ_s into finite pieces or to find a simultaneously μ_s-integrable function $k > 0$ on S and thus to actually prove properness. Conversely if all these efforts fail one might be tempted to guess that properness does not hold. But it seems hard to ensure this without further tools. The inversion kernel κ now actually represents an appropriate tool that will enable us to reject properness in certain cases. Say that a subset $L \subset G$ is *locally closed* if it is the intersection of an open and a closed set. It is well known that such sets inherit local-compactness from G with respect to the inherited topology, see [10, I.65].

Corollary 3.10 (properness and stabilizers)**.** *Let G operate properly on the Borel space S such that $G_{s,s}$ is locally closed in G for all $s \in S$. Then $G_{s,s}$ is compact in G for all $s \in S$.*

Proof. The assumption that $G_{s,s}$ is locally closed in G implies that $G_{s,s}$ is a locally compact subgroup of G and for each s we may choose some left Haar measure λ_s on $G_{s,s}$. Consider the kernel κ from Theorem 3.1. For any $s \in S$, the measure $\kappa_{s,s}$ is concentrated on $G_{s,s}$ and for any $g \in G_{s,s}$ we have by invariance

$$\kappa_{s,s} \circ \theta_g^{-1} = \kappa_{s,gs} = \kappa_{s,s}.$$

Hence $\kappa_{s,s}$ is a left $G_{s,s}$-invariant finite non-zero measure on $G_{s,s}$. The uniqueness result [30, Corollary 2.6] now implies $\lambda_s = c \cdot \kappa_{s,s}$ for some $c \in [0, \infty)$, hence λ_s is finite, which, by a well-known theorem (see e.g. [20, Proposition 11.4 (d)] or [19, Satz 3.15 (b)]), implies compactness of $G_{s,s}$. $\qquad \square$

Examples (non-proper operations)**.** By means of the above Corollary 3.10 it is straightforward to see that $\mathbb{R}^d$ does not operate properly on the Grassmanian $A(k, d)$ via translation. Similarly the operation of $\mathbb{R}^d$ on $\mathcal{F}^d$ - the space of closed subsets of $\mathbb{R}^d$ - via translation is not proper (note that e.g. k-dimensional linear subspaces have closed but non-compact stabilizers).

For a last example consider the group G_d of rigid motions on $\mathbb{R}^d$. It operates transitively on $A(d, k)$ in the canonical way and this operation is, as G_d contains the translations, also not proper. If properness is needed one might instead consider the proper (and also transitive) operation $\mathbb{R}^{d-k} \times SO(d) \hookrightarrow A(d, k)$ defined via $(x, \vartheta, E) \mapsto \vartheta(\psi_{E^\perp}(x) + E)$, where for any $E \in A(d, k)$ the map $\psi_{E^\perp}$ is a fixed vector space isomorphism from $\mathbb{R}^{d-k}$ to $E^\perp$.

3.2.3 Projections of functions on groups

Given a proper operation $G \hookrightarrow S$, our fixed choice of a system of orbit representatives $O \subset S$ (which induces the choice function β) allows for the following canonical transformation of any measurable function f defined on G into a u-measurable function f^* defined on S:

$$f^*(s) := \int f(g^{-1}) \kappa_{\beta(s),s}(dg), \quad s \in S.$$

It maps $s \in S$ to the mean of f on the coset $G_{s,\beta(s)}$. This is particularly convenient if f itself is constant on these cosets such that no functional information on f is lost when forming f^*. An important example is the modular function Δ on G which may be projected to any space S on which it operates properly via

$$\Delta^*(s) := \int \Delta(g^{-1}) \kappa_{\beta(s),s}(dg), \quad s \in S, \tag{3.9}$$

without any loss of information from Δ, since by (2.5) Δ is constant on the cosets of the stabilizers: If $g, h \in G_{\beta(s),s}$ then $g^{-1}h \in G_{\beta(s),\beta(s)}$ so that (2.5) implies $1 = \Delta(g^{-1}h)$, i.e. $\Delta(g^{-1}) = \Delta(h^{-1})$. The important point of the above construction is that it automatically gives u-measurability in s. There are other possible ways for introducing Δ^*:

Lemma 3.12. (Δ^* and Δ) *Let* $G \hookrightarrow S$ *be proper and choose* w *as in* (2.6). *Further for any fixed* $s \in S$ *let* g_s *denote some element of* $G_{\beta(s),s}$. *Then*

$$\Delta^*(s) = \mu_s w = \Delta(g_s^{-1}).$$

Proof. The first equality follows from Fubini's Theorem since

$$\mu_s w = \iint w(gh\beta(s)) \kappa_{\beta(s),s}(dh) \lambda(dg) = \int \Delta(h^{-1}) \kappa_{\beta(s),s}(dh) = \Delta^*(s). \tag{3.10}$$

The other follows from the definition of Δ^* in (3.9) and the fact that Δ is constant on cosets of stabilizers. $\qquad\square$

As seen in various examples in Subsection 3.1.2 and also in the large generality of Corollary 3.10, properness imposes restrictions on the size of stabilizers which lead to explicit formulas when computing Δ^* in special cases. Besides the trivial case where G is unimodular and hence $\Delta^* \equiv 1$ another computable (non-trivial) case is that of countable S which is of independent interest for applications (e.g. for percolation on countable graphs, see [6],[44]), also see Subsection 5.4.2.

Lemma 3.13. (Δ^* for countable S) *If* G *operates properly on a countable set* S *then*

$$\Delta^*(s) = \frac{\lambda(G_{s,s})}{\lambda(G_{\beta(s),\beta(s)})} = \frac{|G_{s,s}\beta(s)|}{|G_{\beta(s),\beta(s)}s|}, \quad s \in S. \tag{3.11}$$

Proof. Choose $k > 0$ on S such that $\sum_{s \in Gb} k(s) = \sum_{s \in Gb'} k(s) < \infty$, $b, b' \in O$. Then it follows that

$$\Delta^*(s) = \frac{\mu_s k}{\mu_{\beta(s)} k} = \frac{\lambda(G_{s,s}) \sum_{t \in Gs} k(t)}{\lambda(G_{\beta(s),\beta(s)}) \sum_{t \in G\beta(s)} k(t)} = \frac{\lambda(G_{s,s})}{\lambda(G_{\beta(s),\beta(s)})}, \quad s \in S.$$

The second equality in (3.11) follows from Lemma 3.8.   □

In other cases the following derivate of the modular function will be useful. Let G operate properly on Borel spaces S and T. We consider the associated uniquely determined inversion kernels κ^S and κ^T. In addition, assuming their existence, we fix in each of these spaces measurable systems O_S resp. O_T of orbit representatives. Then we may define

$$\tilde{\Delta}(s,t) := \iint \frac{\Delta(g)}{\Delta(h)} \kappa^S_{\beta(s),s}(dg) \kappa^T_{\beta(t),t}(dh), \quad s \in S, t \in T. \tag{3.12}$$

Note that $\tilde{\Delta}$ has the properties

$$\tilde{\Delta}(gs, ht) = \frac{\Delta(g)}{\Delta(h)} \tilde{\Delta}(s,t), \quad g, h \in G, s \in S, t \in T,$$

(in particular $\tilde{\Delta}$ is jointly G-invariant) and

$$\tilde{\Delta}(b,c) = 1, \quad b \in O_S, c \in O_T. \tag{3.13}$$

For later considerations it will be important to note the following representation of $\tilde{\Delta}$ using functions w^S on S and w^T on T as in (2.6) as well as the connection to Δ^*:

Lemma 3.14. ($\tilde{\Delta}$ and Δ^*) *Let G operate on the Borel spaces S and T properly. Then*

$$\tilde{\Delta}(s,t) = \frac{\mu_t w^T}{\mu_s w^S} = \frac{\Delta^*(t)}{\Delta^*(s)}, \quad s \in S, t \in T, \tag{3.14}$$

where w^S and w^T are any functions on S and T respectively satisfying (2.6).

Proof. Using (2.6) we may write

$$\tilde{\Delta}(s,t) = \iint \frac{\Delta(g)}{\Delta(h)} \frac{\mu_{\beta(t)} w^T}{\mu_{\beta(s)} w^S} \kappa^S_{\beta(s),s}(dg) \kappa^T_{\beta(t),t}(dh)$$

$$= \iint \frac{\mu_{h\beta(t)} w^T}{\mu_{g\beta(s)} w^S} \kappa^S_{\beta(s),s}(dg) \kappa^T_{\beta(t),t}(dh)$$

where we used (2.4) and the homomorphism property of Δ in the second step. The first equality now follows, the second is then clear by Lemma 3.12.   □

We quickly mention the special form that $\tilde{\Delta}$ takes when S is countable.

Lemma 3.15. ($\tilde{\Delta}$ for countable S) *Assume that the operations $G \hookrightarrow S$ and $G \hookrightarrow T$ are proper and both S and T are countable. Then for any $s \in S$ and $t \in T$*

$$\tilde{\Delta}(s,t) = \frac{\lambda(G_{t,t})}{\lambda(G_{s,s})} \cdot \frac{\lambda(G_{\beta(s),\beta(s)})}{\lambda(G_{\beta(t),\beta(t)})} = \frac{|G_{t,t}\beta(t)|}{|G_{\beta(t),\beta(t)}t|} \cdot \frac{|G_{\beta(s),\beta(s)}s|}{|G_{s,s}\beta(s)|}.$$

Proof. Apply Lemma 3.13 and Lemma 3.14. $\qquad\square$

Remark 3.16 (countable transitive case). Note that when $S = T$ is countable and $G \hookrightarrow S$ is transitive (this is e.g. the case for a countable transitive graph (V, E) where V is the set of vertices, $E \subset V \times V$ the set of edges and $G := \mathrm{Aut}((V,E))$ the group of graph automorphisms operating on V) then

$$\tilde{\Delta}(s,t) = \frac{\lambda(G_{t,t})}{\lambda(G_{s,s})} = \frac{|G_{t,t}s|}{|G_{s,s}t|} \tag{3.15}$$

(use Lemma 3.15 for the first equality and Lemma 3.8 for the second).

Finally the following lemma is sometimes helpful.

Lemma 3.17 (unimodularity). *Let $G \hookrightarrow S$ and $G \hookrightarrow T$ be proper where S, T are Borel, Δ^* defined on S as in (3.9) and $\tilde{\Delta}$ defined on $S \times T$ as in (3.12). Then the following statements are equivalent:*

(i) *G is unimodular,*

(ii) *$\Delta^* \equiv 1$,*

(iii) *$\tilde{\Delta} \equiv 1$.*

Proof. (i) $\Rightarrow$ (ii) follows from (3.9) and (ii) $\Rightarrow$ (iii) from (3.14). Now assume (iii). Then (3.14) implies $\mu_s w^S = 1, s \in S$, and since $G \hookrightarrow S$ is proper this implies (i) by (2.4) and (2.5). $\qquad\square$

3.2.4 Transforming stationary random measures

Given an operation $G \hookrightarrow S$ we write here $\mathcal{I}$ for the invariant σ-algebra on $M(S)$ with respect to the induced operation and for any random measure ξ on S we put

$$\mathcal{I}_\xi := \xi^{-1}(\mathcal{I}) = \{\{\xi \in I\} : I \in \mathcal{I}\},$$

which is a σ-algebra on Ω. Given a random measure η on S and a measurable function $h : \Omega \times S \to [0, \infty)$ with

$$\int h(\theta_g^{-1}\omega, \beta(s))\kappa_{\beta(s),s}(dg) < \infty, \quad \omega \in \Omega, s \in S, \tag{3.16}$$

we may define another random measure (by Lemma 2.21 (i)) ξ on S via

$$\xi(C) := \iint \mathbf{1}_C(s)h(\theta_g^{-1}, \beta(s))\kappa_{\beta(s),s}(dg)\eta(ds). \tag{3.17}$$

We call ξ the *h-transform of η* for given h and η as above. The important feature of this transformation is that if η is G-stationary then ξ inherits the G-stationarity.

Lemma 3.18. (*h-transform preserves stationarity*) *Given any* $h \in (\mathcal{A} \otimes \mathcal{S})_+$ *and a G-stationary random measure* η *on the Borel space* S, *then its h-transform* ξ *is also G-stationary and further* $\mathcal{I}_\xi \subset \mathcal{I}_\eta$.

Proof. As η is G-stationary we have for $l \in G$

$$
\begin{aligned}
\xi(\theta_h, C) &= \iint \mathbf{1}\{s \in C\} h(\theta_g^{-1}\theta_l, \beta(s)) \kappa_{\beta(s),s}(dg) \eta(\theta_l, ds) \\
&= \iint \mathbf{1}\{ls \in C\} h(\theta_g^{-1}\theta_l, \beta(s)) \kappa_{\beta(s),ls}(dg) \eta(ds) \\
&= \iint \mathbf{1}\{ls \in C\} h(\theta_{lg}^{-1}\theta_l, \beta(s)) \kappa_{\beta(s),s}(dg) \eta(ds) \\
&= \iint \mathbf{1}\{ls \in C\} h(\theta_g^{-1}, \beta(s)) \kappa_{\beta(s),s}(dg) \eta(ds)
\end{aligned}
$$

and by definition of ξ this equals $\xi(l^{-1}C)$. Thus ξ is G-stationary. To establish the second assertion note that $\xi = f(\eta)$ where

$$
f : M(S) \to M(S), \quad f(\mu) := \int \mathbf{1}\{s \in \cdot\} h(\theta_g^{-1}, \beta(s)) \kappa_{\beta(s),s}(dg) \mu(ds)
$$

is *G-covariant* in the sense that

$$
f(\theta_g \mu) = \theta_g(f(\mu)), \quad g \in G.
$$

This readily implies that $f^{-1}(I)$ is G-invariant whenever this is true for $I \subset M(S)$. Hence

$$
\{\xi \in I\} = \{f(\eta) \in I\} = \{\eta \in f^{-1}(I)\} \in \mathcal{I}_\eta
$$

for any such I. $\qquad\qquad\qquad\qquad\qquad\qquad\qquad\qquad\qquad\qquad\qquad\qquad\qquad\qquad\square$

A second useful transformation of a G-stationary random measure ξ on S is the following. We define the *G-transform* of ξ as

$$
\hat{\xi} := \iint \mathbf{1}\{(g, \beta(s)) \in \cdot\} \kappa_{\beta(s),s}(dg) \xi(ds).
$$

We show that $\hat{\xi}$ is a random measure. Since ξ is a random measure we may choose $h : \Omega \times S \to (0, \infty)$ such that $\xi(\omega, h(\omega, \cdot)) < \infty$. Then $f(\omega, g, s) := h(\omega, gs)$ is strictly positive and we have

$$
\hat{\xi}_\omega(f_\omega) = \iint h(\omega, g\beta(s)) \kappa_{\beta(s),s}(dg) \xi(\omega, ds) = \xi(\omega, h(\omega, \cdot)) < \infty,
$$

which gives the assertion. It is clear that ξ may be recovered from $\hat{\xi}$ as its image under $\pi : G \times O \to S, (g, b) \mapsto gb$. It is also evident from the construction that ξ is G-stationary if and only if $\hat{\xi}$ is G-stationary with respect to the operation $G \hookrightarrow G \times O$ given by $(h, (g, b)) \mapsto (hg, b)$. If $G \hookrightarrow S$ is transitive, then O consists of one element and $G \times O$ may be identified with G. Hence in this important special case the G-transform turns G-stationary random measures on S into G-stationary random measures on G. Last [38] used this special form of the above introduced G-transform to extend Palm Calculus from stationary random measures on groups to stationary random measures on homogeneous spaces.

3.2.5 Invariant Palm kernels

Given operations $G \hookrightarrow S$ and $G \hookrightarrow T$ on Borel spaces S and T together with a random measure ξ on S and a random element η in T such that (ξ, η) is a *jointly G-stationary pair* in the sense that

$$(g\xi, g\eta) \stackrel{d}{=} (\xi, \eta), \quad g \in G,$$

it is straightforward to check that the Campbell measure

$$C_{\xi,\eta} f = \mathbb{E} \int f(\eta, s)\xi(ds), \quad f \in (\mathcal{T} \otimes \mathcal{S})_+,$$

of (ξ, η) is jointly G-invariant. As it is s-finite by Lemma 2.19 and since T is Borel, Theorem 3.9 yields invariant disintegrations of the form

$$C_{\xi,\eta} = \iint f(t, s) P_s(dt)\nu(ds), \quad f \in (\mathcal{T} \otimes \mathcal{S})_+. \tag{3.18}$$

Also, given a fixed σ-finite G-invariant measure $\tilde{\nu} \gg M(\cdot \times T)$ on S Theorem 3.9 yields a suitable invariant kernel of Palm pseudo distributions $\tilde{P}_s$ (*pseudo* refers to the fact that the P_s may not be stochastic) such that

$$C_{\xi,\eta} = \iint f(t, s)\tilde{P}_s(dt)\tilde{\nu}(ds), \quad f \in (\mathcal{T} \otimes \mathcal{S})_+. \tag{3.19}$$

Let us recall that Theorem 3.9 also gives that $C_{\xi,\eta}$ and P are simultaneously σ-finite and that Lemma 2.19 (iii) contains sufficient conditions for this. Hence it is always possible to choose invariant versions of our Palm pseudo distributions of η with respect to ξ. These insights (besides our slight extension to s-finite Campbell measures) all stem in this generality from [30] and go back to [61, 45].

Chapter 4

Palm Theory

Classically, Palm theory was developed for completely stationary random measures on $\mathbb{R}^d$, i.e. random measures stationary with respect to all translations. In our language this means stationarity with respect to the operation of $\mathbb{R}^d$ on itself via translation, which we shall always denote by $\mathbb{R}^d \hookrightarrow \mathbb{R}^d$. As explained in Subsection 2.4.2 we may model a stationary random measure ξ without loss of generality by assuming the existence of an abstract measurable flow $\{\theta_x : x \in \mathbb{R}^d\}$ on Ω, requiring the underlying measure $\mathbb{P}$ on $(\Omega, \mathcal{A})$ to be invariant with respect to shifts induced by this flow and by requiring ξ to be adapted to this flow in the sense that

$$\xi(\theta_x \omega, x + A) = \xi(\omega, A), \quad \omega \in \Omega, A \in \mathcal{B}^d(\mathbb{R}^d). \tag{4.1}$$

Given such a $\mathbb{R}^d$-stationary random measure on $\mathbb{R}^d$, its Palm measure $\mathbb{Q}$ on Ω with respect to $O = \{0\}$ is defined as

$$\mathbb{Q}(\cdot) = \mathbb{E} \int \mathbf{1} \left\{ \theta_x^{-1} \in \cdot \right\} \mathbf{1}_{[0,1]^d}(x) \xi(dx).$$

Evidently it combines a spatial averaging over $[0,1]^d$ with respect to ξ and a phase averaging over Ω with respect to $\mathbb{P}$. If $\xi \neq 0$ has *finite intensity*

$$0 < \gamma_\xi := \mathbb{E}\xi[0,1]^d = \mathbb{Q}(\Omega) < \infty,$$

we may normalize $\mathbb{Q}$ to a probability measure $\mathbb{P}_0$ via

$$\mathbb{P}_0(\cdot) = \frac{1}{\gamma_\xi} \mathbb{Q}(\cdot).$$

This measure plays a prominent role in Stochastic Geometry, since it allows the extraction of meaningful distributions of objects derived from spatially unbounded, stationary processes. Examples are the notions of *typical grain* of a stationary particle process, *typical cell* of a stationary tessellation or partition, or the *directional distribution* of a k-flat process. All these objects may be interpreted as the distributions of suitable random objects under the Palm probability measure. In the cases of G-stationary random subgraphs, tessellations and partitions, we shall construct analogues objects even for non-transitive operations using the tools developed in this chapter later in Chapters 5 and 7.

If ξ is a simple point process, then clearly $\mathbb{Q}$, and thus $\mathbb{P}_0$, is concentrated on all configurations $\omega \in \Omega$ having $\xi(\omega, \{0\}) = 1$. In fact, in this case $\mathbb{P}_0$ may be interpreted as the conditional probability derived from $\mathbb{P}$, given the information that ξ has a point in the origin, see [28, Theorem 11.5]. Thus, under the Palm measure, ξ has a point at the origin and this point in the origin then receives the interpretation of a *typical point* of ξ. It should be mentioned now, that the ambiguous word *typical* leaves space for interpretations and in fact there are other possibilities to formalize other ideas for *typical*. As should be clear from its definition, the Palm measure favors configurations ξ with a high sample intensity

$$\hat{\xi} = \lim_{r \to \infty} \frac{\xi(rB)}{\lambda^d(rB)},$$

and thus the word *typical*, even though often used in an informal manner in many places in the relevant literature, has to be read with great care. Under ergodicity assumptions on ξ, this sample intensity is constant, and the interpretation of the origin as a typical point of ξ is more accurate. In any case, it is interesting to inspect derived random objects from a stationary random measure under this Palm measure. We shall do so in Chapter 7.

Here, in Chapter 4, we shall introduce an analogue of the above Palm measure $\mathbb{Q}$ for a random measure ξ on a measurable space S, on which a general lcsc group G acts in some way. We neither require G to be unimodular nor the action $G \hookrightarrow S$ to be transitive. This analogue, the *cumulative Palm measure* will be constructed in Section 4.1. We shall then proceed in Section 4.2 by explicitly computing the cumulative Palm measure for G-stationary Cox processes and even characterize G-stationary Cox processes in terms of their cumulative Palm measures in a Slivnyak-type manner [65], cf. [29, Corollary 2.35]. We then proceed in Section 4.3 with an illustration on how probability measures may be derived from the cumulative Palm measure and how these probability measures relate to each other via conditioning. The final Section 4.4 contains two main results about cumulative Palm measures, namely a Neveu exchange type formula for transport kernels (referred to as *transport theorem*) and an intrinsic characterization of a cumulative Palm measure of ξ (generalizing a result of Mecke [46]), both independently proved in [21] and Kallenberg [31]. In this section, we also characterize *balancing weighted transport kernels* between random measures in terms of a transport equation between their cumulative Palm measures.

4.1 The cumulative Palm measure

We consider a lcsc group G with Haar measure λ operating properly on a Borel space $(S, \mathcal{S})$ together with a random measure ξ on S which is G-stationary. We are not assuming transitivity of the group operation. In Subsection 4.1.1 we introduce the fundamental object of interest in this general mathematical frame: the *cumulative Palm measure* of ξ.

4.1.1 Construction

In this section we will construct and define a central object of this thesis. Recall the function w from (2.6) which exists if $G \hookrightarrow S$ is proper. The following theorem makes crucial use of the inversion kernel and is even new in the transitive classical setting $\mathbb{R}^d \hookrightarrow \mathbb{R}^d$ (even though the inversion kernel doesn't play a real role in this setting since trivially $\kappa_{\beta(s),s} = \kappa_{0,s} = \delta_s, s \in \mathbb{R}^d$). Assume that ξ is a G-stationary random measure on S. Then it is easy to see that the Campbell measure of ξ is jointly G-invariant. If $(\Omega, \mathcal{A})$ is Borel, Theorem 2.16 (see also Corollary 3.5 in [30]) implies that there is a G-invariant Palm pair (ν, Q) of ξ, meaning that both ν and Q are G-invariant. In probability theory it is not common to put extra technical requirements upon the underlying space $(\Omega, \mathcal{A})$. As we will see in the following Theorem 4.1, there is a single and more natural object to look at when investigating partially stationary random measures. Its construction does not depend on Palm pairs and its existence may be derived without technical assumptions on Ω.

Theorem 4.1 (existence and uniqueness). *Let G operate properly on the Borel space S and fix a system of representatives $O \in \mathcal{S}$. Further let ξ be a G-stationary random measure on S. Then there is a unique measure $\mathbb{Q}$ on $\Omega \times S$ concentrated on $\Omega \times O$ satisfying both the* refined Campbell equation

$$\mathbb{E} \int f(\theta_e, s)\xi(ds) = \iint f(\theta_g\omega, gb)\lambda(dg)\mathbb{Q}(d(\omega, b)), \quad f \in (\mathcal{A} \otimes \mathcal{S})_+, \qquad (4.2)$$

and the stabilizer invariance condition

$$\iint f(\theta_h^{-1}\omega, b)\kappa_{b,b}(dh)\mathbb{Q}(d(\omega, b)) = \int f(\omega, b)\mathbb{Q}(d(\omega, b)), \quad f \in (\mathcal{A} \otimes \mathcal{S})_+. \qquad (4.3)$$

In addition $\mathbb{Q}$ is σ-finite and for any function $w : S \to [0, \infty)$ with $\mu_b w = 1, b \in O$, it is given by

$$\mathbb{Q}(\cdot) = \mathbb{E} \iint \mathbf{1}\left\{\left(\theta_g^{-1}, \beta(s)\right) \in \cdot\right\}\kappa_{\beta(s),s}(dg)w(s)\xi(ds). \qquad (4.4)$$

Proof. To prove the existence of $\mathbb{Q}$, we fix an arbitrary $w : S \to [0, \infty)$ satisfying $\mu_b w = 1, b \in O$, e.g. by means of Lemma 2.6 (as we may since $G \hookrightarrow S$ is proper). Then we define $\mathbb{Q}$ as in (4.4) and compute

$$\iint f(\theta_g\omega, gb)\lambda(dg)\mathbb{Q}(d(\omega, b)) = \mathbb{E} \iiint f(\theta_g\theta_h^{-1}, g\beta(s))\lambda(dg)\kappa_{\beta(s),s}(dh)w(s)\xi(ds)$$

$$= \mathbb{E} \iiint f(\theta_{gh^{-1}}, gh^{-1}s)\lambda(dg)\kappa_{\beta(s),s}(dh)w(s)\xi(ds)$$

$$= \mathbb{E} \iiint \Delta(h)f(\theta_g, gs)\lambda(dg)\kappa_{\beta(s),s}(dh)w(s)\xi(ds).$$

Using Fubini, G-invariance of $\mathbb{P}$ and G-stationarity of ξ we arrive at

$$\iint f(\theta_g\omega, gb)\lambda(dg)\mathbb{Q}(d(\omega, b)) = \int \mathbb{E} \iint \Delta(h)f(\theta_e, s)w(g^{-1}s)\kappa_{\beta(s),g^{-1}s}(dh)\xi(ds)\lambda(dg)$$

$$= \mathbb{E} \iiint \Delta(g^{-1}h)w(g^{-1}s)\lambda(dg)\kappa_{\beta(s),s}(dh)f(\theta_e, s)\xi(ds).$$

Then, using property (2.2) of the modular function we proceed as follows:

$$\iint f(\theta_g\omega, gb)\lambda(dg)\mathbb{Q}(d(\omega, b)) = \mathbb{E} \iiint \Delta(h)w(gs)\lambda(dg)\kappa_{\beta(s),s}(dh)f(\theta_e, s)\xi(ds)$$

$$= \mathbb{E} \iiint w(gh^{-1}s)\lambda(dg)\kappa_{\beta(s),s}(dh)f(\theta_e, s)\xi(ds)$$

$$= \mathbb{E} \iint w(g\beta(s))\lambda(dg)f(\theta_e, s)\xi(ds)$$

$$= \mathbb{E} \int f(\theta_e, s)\xi(ds).$$

Hence this $\mathbb{Q}$ satisfies (4.2). It also satisfies (4.3) since the left-hand side of (4.3) equals

$$\mathbb{E} \iiint f(\theta_h^{-1}\theta_g^{-1}, \beta(s))\kappa_{\beta(s),\beta(s)}(dh)\kappa_{\beta(s),s}(dg)w(s)\xi(ds)$$

$$= \mathbb{E} \iiint f(\theta_{gh}^{-1}, \beta(s))\kappa_{\beta(s),\beta(s)}(dh)\kappa_{\beta(s),s}(dg)w(s)\xi(ds)$$

$$= \mathbb{E} \iiint f(\theta_h^{-1}, \beta(s))\kappa_{\beta(s),g\beta(s)}(dh)\kappa_{\beta(s),s}(dg)w(s)\xi(ds)$$

$$= \mathbb{E} \iint f(\theta_h^{-1}, \beta(s))\kappa_{\beta(s),s}(dh)w(s)\xi(ds)$$

$$= \mathbb{Q}f.$$

To establish uniqueness let $\mathbb{Q}$ and $\tilde{\mathbb{Q}}$ satisfy both (4.2) and (4.3). Then in particular

$$\iint f(\theta_g\omega, gb)\lambda(dg)\mathbb{Q}(d(\omega,b)) = \iint f(\theta_g\omega, gb)\lambda(dg)\tilde{\mathbb{Q}}(d(\omega,b)), \quad f \in (\mathcal{A} \otimes \mathcal{S})_+,$$

and choosing $f(\omega, s) := w(s) \int \tilde{f}(\theta_h^{-1}\omega, \beta(s))\kappa_{\beta(s),s}(dh)$ for arbitrary $\tilde{f} \in (\mathcal{A} \otimes \mathcal{S})_+$ this becomes

$$\iiint \tilde{f}(\theta_h^{-1}\theta_g\omega, b)\kappa_{b,gb}(dh)w(gb)\lambda(dg)\mathbb{Q}(d(\omega,b))$$

$$= \iiint \tilde{f}(\theta_h^{-1}\theta_g\omega, b)\kappa_{b,gb}(dh)w(gb)\lambda(dg)\tilde{\mathbb{Q}}(d(\omega,b)),$$

which means (since $\int w(gb)\lambda(dg) = 1$)

$$\iint \tilde{f}(\theta_h^{-1}\omega, b)\kappa_{b,b}(dh)\mathbb{Q}(d(\omega,b)) = \iint \tilde{f}(\theta_h^{-1}\omega, b)\kappa_{b,b}(dh)\tilde{\mathbb{Q}}(d(\omega,b)),$$

and hence $\mathbb{Q} = \tilde{\mathbb{Q}}$ by (4.3). Finally we note that $\mathbb{Q}$ must be σ-finite: Since C_ξ is σ-finite we may choose a measurable function $f : \Omega \times S \to (0,\infty)$ such that $C_\xi f < \infty$. Then

$$g(\omega, b) := \int f(\theta_g\omega, gb)\lambda(dg), \quad \omega \in \Omega, b \in O,$$

is strictly positive, and by (4.2) $\mathbb{Q}g = C_\xi f < \infty$. $\square$

This theorem gives rise to the following definition:

Definition 4.2 (cumulative Palm measure). Given a random measure ξ on the Borel space S the unique measure $\mathbb{Q}$ satisfying both (4.2) and (4.3) in Theorem 4.1 is called the *cumulative Palm measure* of ξ with respect to O. We may sometimes write $\mathbb{Q}^\xi := \mathbb{Q}$ to make the dependence on ξ explicit.

The word *cumulative* indicates the fact that $\mathbb{Q}$ is a superposition of ordinary Palm measures, as we shall see in Theorems 4.10 and 4.12. Instead of using the function w from Lemma 2.9 we may clearly use G-symmetric sets instead whenever they exist.

Corollary 4.3 (cumulative Palm measure and symmetric sets). *If B is a G-symmetric set then*

$$\mathbb{Q}(\cdot) = \frac{1}{\delta(B)}\mathbb{E} \iint \mathbf{1}\{(\theta_g^{-1}, \beta(s)) \in \cdot\}\kappa_{\beta(s),s}(dg)\mathbf{1}_B(s)\xi(ds). \tag{4.5}$$

Proof. It is enough to note that $\tilde{w}(s) := \mathbf{1}_B(s)/\delta(B)$ has the property $\mu_b \tilde{w} = 1, b \in O$. $\qquad\square$

Example 4.4 (classical setting). Considering the classical setting $\mathbb{R}^d \hookrightarrow \mathbb{R}^d$, equation (4.3) is trivially always fulfilled and may be omitted. Choosing $O = \{0\}$ and identifying $\Omega \times \{0\}$ with Ω the refined Campbell equation reduces to

$$\mathbb{E} \int f(\theta_e, s)\xi(ds) = \iint f(\theta_g \omega, g)\lambda(dg)\mathbb{Q}(d\omega), \quad f \in (\mathcal{A} \otimes \mathcal{S})_+.$$

Hence the cumulative Palm measure $\mathbb{Q}$ with respect to ξ and $O = \{0\}$ is nothing but the ordinary Palm measure of ξ at 0 (see [14, 15, 38, 28, 29]).

Example 4.5 (transitive setting). Considering the transitive situation as in Example 3.3 we may fix some $c \in S$ which serves as representative for the single orbit, i.e. $O = \{c\}$ and $\beta \equiv c$. Then the cumulative Palm measure of a G-stationary random measure ξ on S with respect to $\{c\}$ is given by

$$\mathbb{Q}(\cdot) = \mathbb{E} \iint \mathbf{1}\{(\theta_g^{-1}, c) \in \cdot\}\kappa_{c,s}(dg)w(s)\xi(ds).$$

Identifying $\Omega \times \{c\}$ with Ω, this is exactly the ordinary Palm measure of stationary random measures on homogenous spaces, compare e.g. in [38] the equations (3.12), (3.8) and (3.5). In the further special case $S = G$ of Example 3.4 we may take $c = e$ and the cumulative Palm measure further simplifies to

$$\mathbb{Q}(\cdot) = \mathbb{E} \int \mathbf{1}\{(\theta_g^{-1}, e) \in \cdot\}w(g)\xi(dg). \tag{4.6}$$

Under the same identification as above this is the ordinary Palm measure for random measures on groups, see e.g. [46], [39] or [68].

Example 4.6 (completely non-stationary setting). As noted in Example 3.5 the case where no stationarity or invariance assumptions are made may be treated as a special case of our framework, namely by choosing $G = \{e\}$ where $\lambda = \delta_e$, $O = S$ and for any $s \in S$

$$\beta(s) = s, \quad \mu_s = \delta_s, \quad \text{and} \quad \kappa_{\beta(s),s} = \kappa_{s,s} = \delta_e.$$

First, note that (4.3) reduces to a condition which is always satisfied. In view of (4.2) it is then clear that

$$\mathbb{Q} = C_\xi,$$

i.e. $\mathbb{Q}$ is the Campbell measure of ξ in this case. Since the only possible choice for w as in (2.6) is given by $w \equiv 1$ in this situation this is also consistent with (4.4).

Example 4.7 (cumulative Palm measure of deterministic measures). A deterministic measure ν on a measurable space S is G-stationary with respect to an operating group G if and only if it is G-invariant. We now compute the cumulative Palm measure of a G-invariant measure ν on S. By (4.4) we have

$$\mathbb{Q}^\nu = \mathbb{E} \iint \mathbf{1}\{(\theta_g^{-1}, \beta(s)) \in \cdot\}\kappa_{\beta(s),s}(dg)w(s)\nu(ds)$$

and Fubini and G-invariance of $\mathbb{P}$ yield

$$\mathbb{Q}^\nu = \int \mathbb{P}((\theta_e, \beta(s)) \in \cdot)w(s)\nu(ds).$$

Using the orbital decomposition of ν from (2.7) yields

$$\mathbb{Q}^\nu = \mathbb{P} \otimes \nu^*. \tag{4.7}$$

4.1.2   Basic properties

It is important to note a basic support property of the cumulative Palm measure in the case when ξ is a point process:

Lemma 4.8 (concentration property). *Let $\mathbb{Q}$ be the cumulative Palm measure of a G-stationary point process ξ on a Borel space S with respect to a fixed measurable system of orbit representatives O. Then*

$$\mathbb{Q}(\{(\omega, b) \in \Omega \times O : \xi(\omega, \{b\}) = 0\}) = 0.$$

Proof. By (4.4) the left side equals

$$\mathbb{E} \iint \mathbf{1}\{\xi(\theta_g^{-1}\omega, \{\beta(s)\}) = 0\}\kappa_{\beta(s),s}(dg)w(s)\xi(ds)$$

which by (2.27) may be written as

$$\mathbb{E} \iint \mathbf{1}\{\xi(\omega, \{g\beta(s)\}) = 0\}\kappa_{\beta(s),s}(dg)w(s)\xi(ds) = \mathbb{E} \int \mathbf{1}\{\xi(\omega, \{s\}) = 0\}w(s)\xi(ds).$$

This last expression is clearly 0 as ξ is a point process. $\qquad\qquad\square$

We emphasize here that the cumulative Palm measure $\mathbb{Q}$ of a random measure ξ on a Borel space S with respect to a system O of orbit representatives in S emerged by factoring out the Haar measure λ from the Campbell measure C_ξ of ξ, see (4.2). As C_ξ clearly contains all information about $\mathbb{P}$ outside of $\mathbf{1}\{\xi = 0\}$, the same must be true for $\mathbb{Q}$, and the following lemma makes this precise. In the special case of random measures on groups it reduces to a formula found by Mecke [46], while for random measures on a homogeneous space the formula can be found in [60].

Lemma 4.9 (inversion formula). *Given a random measure ξ on a Borel space S the underlying measure $\mathbb{P}$ on Ω may be reconstructed from the cumulative Palm measure $\mathbb{Q}$ of ξ with respect to a fixed measurable system O of orbit representatives via*

$$\mathbb{E}[f \cdot \mathbf{1}\{\xi \neq 0\}] = \iint f(\theta_g\omega)h(\theta_g\omega, gb)\lambda(dg)\mathbb{Q}(d(\omega, b)), \quad f \in \mathcal{A}_+,$$

where h is a fixed measurable function $h : \Omega \times S \to (0, \infty)$ satisfying

$$\int h(\omega, s)\xi(\omega, ds) = \mathbf{1}\{\xi(\omega) \neq 0\}, \quad \omega \in \Omega.$$

Proof. Choosing h as in Lemma 2.22 (i) and replacing $f(\omega, s)$ in (4.2) by $\tilde{f}(\omega)h(\omega, s)$ for arbitrary $\tilde{f} \in \mathcal{A}_+$ yields the assertion. $\qquad\square$

The earlier announced connection between $\mathbb{Q}$ and any possibly existing Palm pair (ν, Q) is the following. Here, $\mathbf{b} : O \to O$ denotes the identity map on O, i.e. $\mathbf{b}(b) = b, b \in O$. Recall the *-operator from (2.7).

Theorem 4.10 (link to Palm pairs). *Let ξ be a G-stationary random measure on the Borel space S and let $G \hookrightarrow S$ be proper. If a G-invariant Palm pair (ν, Q) of ξ exists, then it is related to the cumulative Palm measure $\mathbb{Q}$ of ξ via*

$$\mathbb{Q}(\cdot) = \iint \mathbf{1}\{(\omega, b) \in \cdot\}Q_b(d\omega)\nu^*(db). \tag{4.8}$$

Proof. Assume a G-invariant Palm pair (ν, Q) of ξ is given. Then we may define $\bar{\mathbb{Q}} := \nu^* \otimes Q$ and note that

$$
\begin{aligned}
\mathbb{E}\int f(\theta_e, s)\xi(ds) &= \iint f(\omega, s)Q_s(d\omega)\nu(ds) \\
&= \iiint f(\omega, s)Q_s(d\omega)\mu_b(ds)\nu^*(db) \\
&= \iiint f(\theta_g\omega, gb)\lambda(dg)Q_b(d\omega)\nu^*(db) \\
&= \iint f(\theta_g\omega, gb)\lambda(dg)\bar{\mathbb{Q}}(d(\omega, b)).
\end{aligned}
$$

Thus $\bar{\mathbb{Q}}$ satisfies (4.2). It also satisfies (4.3) since by Fubini's theorem

$$
\begin{aligned}
\iint f(\theta_h^{-1}\omega, b)\kappa_{b,b}(dh)\bar{\mathbb{Q}}(d(\omega, b)) &= \iiint f(\theta_h^{-1}\omega, b)Q_b(d\omega)\kappa_{b,b}(dh)\nu^*(db) \\
&= \iiint f(\omega, b)Q_{h^{-1}b}(d\omega)\kappa_{b,b}(dh)\nu^*(db) \\
&= \iint f(\omega, b)Q_b(d\omega)\nu^*(db) \\
&= \int f(\omega, b)\bar{\mathbb{Q}}(d(\omega, b)).
\end{aligned}
$$

The uniqueness assertion in Theorem 4.1 now implies equation (4.8). $\qquad\square$

The use of this theorem hinges on the existence of Palm pairs. These exist e.g. if Ω is Borel. But in the important special case when S is countable they exist even without the Borel assumption upon Ω. For convenience we formulate this result only for simple point processes on S even though (partial) extensions to random measures with σ-finite intensity measure are not hard to obtain. We identify simple point processes with their support.

Theorem 4.11. (Palm pairs for countable S) *Let ξ be a G-stationary simple point process on the countable space S with $\mathbb{P}(s \in \xi) > 0, s \in V$, where $G \hookrightarrow S$ is proper. Then*

$$(\mathbb{E}\xi, (\mathbb{P}(\,\cdot\,|s \in \xi))_{s \in S})$$

is a G-invariant Palm pair of ξ, and fixing a complete system of orbit representatives O

$$(\mathbb{E}\xi)^* = \sum_{b \in O} \frac{\mathbb{P}(b \in \xi)}{\lambda(G_{b,b})}\delta_b. \tag{4.9}$$

Proof. Putting $P(s, \cdot) := \mathbb{P}(\,\cdot\,|\xi\{s\} = 1)$ we have for $C \in \mathcal{A}$

$$P(gs, \theta_g C) = \mathbb{P}(\theta_g C | \xi\{gs\} = 1) = \frac{\mathbb{P}(\theta_g C \cap \{\xi\{gs\} = 1\})}{\mathbb{P}(\xi\{gs\} = 1)}.$$

Here $\{\xi(gs) = 1\} = \theta_g\{\xi(s) = 1\}$ which implies invariance of P. Further for $A \in \mathcal{A}$ and $B \subset S$

$$(\mathbb{E}\xi \otimes P)(B \times A) = \sum_{s \in S} \mathbb{E}\xi\{s\} \frac{\mathbb{P}(A \cap \{\xi\{s\} = 1\})}{\mathbb{P}(\xi\{s\} = 1)} \mathbf{1}\{s \in B\}.$$

As ξ is simple we have $\mathbb{E}\xi\{s\} = \mathbb{P}(\xi\{s\} = 1)$ and since $\mathbf{1}\{\xi\{s\} = 1\} = \xi\{s\}$ we conclude

$$(\mathbb{E}\xi \otimes P)(B \times A) = \sum_{s \in S} \mathbb{P}(A \cap \{\xi\{s\} = 1\}) \mathbf{1}\{s \in B\}$$

$$= \int \sum_{s \in S} \mathbf{1}\{s \in B\} \mathbf{1}\{\omega \in A\} \xi(\omega, \{s\}) \mathbb{P}(d\omega)$$

$$= \mathbb{E} \int \mathbf{1}\{(\theta_e, s) \in A \times B\} \xi(ds).$$

A monotone class argument yields $\mathbb{E}\xi \otimes P = C_\xi$ and thus $(\mathbb{E}\xi, P)$ is a Palm pair of ξ. Equation (4.9) readily follows from (2.11), since ξ is simple. $\qquad\square$

It is clear that combining Theorems 4.10 and 4.11 yields for a simple point process ξ on a countable space S the explicit formula

$$\mathbb{Q}^\xi(\cdot) = \sum_{b \in O} \frac{\mathbb{P}(b \in \xi)}{\lambda(G_{b,b})} \mathbb{P}((\theta_e, b) \in \cdot | \xi\{b\} = 1). \tag{4.10}$$

The next theorem shows how certain important pushforwards of the cumulative Palm measure are related to the (always existent!) invariant disintegrations as in (3.18). We denote the identity map on O by $\mathbf{b}$, i.e. $\mathbf{b}(b) = b, b \in O$.

Theorem 4.12 (pushforwards of the cumulative Palm measure). *Let ξ be a random measure on the Borel space S and η a random element in the Borel space T such that (ξ, η) is jointly G-stationary. Then any invariant disintegration (ν, P) of $C_{\xi,\eta}$ as in (3.18) satisfies*

$$\mathbb{Q}^\xi((\eta, \mathbf{b}) \in \cdot) = \nu^* \otimes P.$$

Proof. By (4.4) we have

$$\mathbb{Q}^\xi((\eta, \mathbf{b}) \in \cdot) = \mathbb{E} \iint \mathbf{1}\left\{\left(\eta(\theta_g^{-1}), \beta(s)\right) \in \cdot\right\} \kappa_{\beta(s),s}(dg) w(s) \xi(ds)$$

$$= \mathbb{E} \iint \mathbf{1}\left\{\left(g^{-1}\eta, \beta(s)\right) \in \cdot\right\} \kappa_{\beta(s),s}(dg) w(s) \xi(ds).$$

Applying (3.18) yields

$$\mathbb{Q}^\xi((\eta, \mathbf{b}) \in \cdot) = \iiint \mathbf{1}\left\{\left(g^{-1}t, \beta(s)\right) \in \cdot\right\} \kappa_{\beta(s),s}(dg) w(s) P_s(dt) \nu(ds)$$

$$= \iiint \mathbf{1}\left\{(t, \beta(s)) \in \cdot\right\} P_{g^{-1}s}(dt) \kappa_{\beta(s),s}(dg) w(s) \nu(ds)$$

where we used Fubini and G-invariance of P. But this clearly equals

$$\iint \mathbf{1}\left\{(t,\beta(s)) \in \cdot\right\} P_{\beta(s)}(dt)w(s)\nu(ds) = \iiint \mathbf{1}\left\{(t,b) \in \cdot\right\} P_b(dt)w(s)\mu_b(ds)\nu^*(db)$$

where we used the decomposition of ν from (2.7). By (2.6) this yields the assertion. $\square$

Slightly rewriting the first of its two defining equations (4.2) and combining it with the second (4.3) yields the following identity for the cumulative Palm measure of a random measure ξ, which will be useful later:

Lemma 4.13 (refined Campbell formula). *For a G-stationary random measure ξ on S the following holds:*

$$\mathbb{E}\iint f(\theta_g^{-1},g,\beta(s))\kappa_{\beta(s),s}(dg)\xi(ds) = \iint f(\omega,g,b)\lambda(dg)\mathbb{Q}(d(\omega,b)). \qquad (4.11)$$

Proof. Starting on the right side we have by (4.3), Fubini and right $G_{b,b}$-invariance of λ

$$\iint f(\omega,g,b)\lambda(dg)\mathbb{Q}(d(\omega,b)) = \iiint f(\theta_h^{-1}\omega,gh,b)\kappa_{b,b}(dh)\lambda(dg)\mathbb{Q}(d(\omega,b)).$$

Using invariance of κ we arrive at

$$\iiint f(\theta_h^{-1}\theta_g\omega,h,b)\kappa_{b,gb}(dh)\lambda(dg)\mathbb{Q}(d(\omega,b))$$

and finally by (4.2) at the left side of the assertion. $\square$

Example 4.14 (intensity measure of the h-transform). Given a G-stationary random measure η on S and a measurable function $h \in (\mathcal{A} \otimes \mathcal{S})_+$ we recall the h-transform ξ of η defined in (3.17) via

$$\xi(C) := \iint \mathbf{1}\{s \in C\}h(\theta_g^{-1},\beta(s))\kappa_{\beta(s),s}(dg)\eta(ds).$$

Lemma 4.13 reveals its intensity measure, since

$$\begin{aligned}
\mathbb{E}\xi(C) &= \mathbb{E}\iint \mathbf{1}\{g\beta(s) \in C\}h(\theta_g^{-1},\beta(s))\kappa_{\beta(s),s}(dg)\eta(ds) \\
&= \iint \mathbf{1}\{gb \in C\}h(\omega,b)\lambda(dg)\mathbb{Q}^\eta(d(\omega,b)) \\
&= \int \mu_b(C)h(\omega,b)\mathbb{Q}^\eta(d(\omega,b)).
\end{aligned}$$

4.2   Palm measure of Cox processes

The aims of this section are first to present a multivariate Slivnyak-Mecke-type formula for Cox process in Subsection 4.2.2, to calculate the cumulative Palm measure of a G-stationary Cox process on an arbitrary space S in Subsection 4.2.3 and to characterize G-stationary Cox processes in terms of their cumulative Palm measure in the spirit of Slivnyak's [65] famous theorem for Poisson processes (see e.g. [29, Corollary 2.35] and also [34, 35]). As Cox processes are (important) generalizations of Poisson processes the calculation of the cumulative Palm measures of G-stationary Poisson processes is included thereby. In Subsection 4.2.1 we first summarize well-known properties of Poisson processes and extend these in Subsection 4.2.2 to Cox processes. Here Theorem 4.18 characterizes Cox processes in terms of equation (4.14). The multivariate Cox Formula in Theorem 4.19 generalizes the multivariate Mecke-Slivnyak formula for Poisson processes and slightly generalizes results of Kallenberg in [33, Theorem 4.2], see Remark 4.20.

4.2.1   Some classical results for Poisson processes

Given a measurable space S and a σ-finite measure μ on S, a *Poisson process* on S *based on* μ is a random measure ξ on S such that for any $n \in \mathbb{N}$ and disjoint $B_1, ..., B_n \in \mathcal{S}$ the random variables $\xi(B_1), ..., \xi(B_n)$ are independent (this property is often paraphrased by saying that ξ has *independent increments*, a term motivated by the 1-dimensional situation of a Poisson process on the line) and such that for any $B \in \mathcal{S}$ the random variable $\xi(B)$ is Poisson distributed with mean $\mu(B)$ if $\mu(B) < \infty$. As μ is σ-finite there is a partition of S into measurable sets $B_1, B_2, \ldots$ such that $\mu(B_i) < \infty$. Putting $P := \{B_1, B_2, ...\}$ we have $\mathbb{P}(\xi \in \mathbf{M}^P(S)) = 1$ where $\mathbf{M}^P(S)$ is defined as in (2.23). As mentioned in Subsection 2.4.1, $\mathbf{M}^P(S)$ is Borel whenever S is which we shall assume in this section, such that we may interpret ξ as a random element in a Borel space.

It is well-known that a 'completely' stationary (i.e. homogeneous) Poisson process in $\mathbb{R}^d$ with finite intensity possesses a very simple *Palm distribution* (its distribution under its Palm probability measure at 0), namely the distribution of $\xi + \delta_0$ under $\mathbb{P}$. Conversely a stationary point process in $\mathbb{R}^d$ with finite intensity whose Palm distribution at 0 is given by $\mathbb{P}(\xi + \delta_0 \in \cdot)$ is a stationary Poisson process. This equivalence is also known as Slivnyak's Theorem, see [63, Satz 3.3.6] for the first stated implication. Slivnyak's result may be stated similarly without stationarity assumptions by invoking the complete kernel of Palm distributions (see (2.24)) instead. A proof of the following result may be found in [29, Theorem 2.34].

Theorem 4.15 (Poisson criterion, Slivnyak). *A point process ξ on a Borel space S with σ-finite intensity measure is Poisson if and only if its Palm distributions $\mathbb{P}(\xi \in \cdot \| \xi)_s$ (defined via (2.24)) are given by $\mathbb{P}(\xi + \delta_s \in \cdot)$ for $\mathbb{E}\xi$-a.e. $s \in S$.*

Note that ξ is neither required to be simple nor that $\mathbb{E}\xi$ is atom-free. This criterion may be equivalently stated in form of the *Slivnyak-Mecke-equation* (4.12) due to Mecke [46, Satz 3.1]. It constitutes an integrated version of Theorem 4.15 and the derivation of either theorem from the other is a trivial consequence of (2.24).

Theorem 4.16 (Poisson criterion, Mecke). *A point process ξ on a Borel space S with σ-finite intensity measure is Poisson if and only if*

$$\mathbb{E} \int f(\xi, s)\xi(ds) = \int \mathbb{E}f(\xi + \delta_s, s)(\mathbb{E}\xi)(ds) \qquad (4.12)$$

for all measurable functions $f : \mathbf{M}(S) \times S \to [0, \infty]$.

Theorem 4.16 even holds without Borel assumption on S, see [46, Satz 3.1]. It is this beautiful criterion that leads to the fundamental and in Stochastic Geometry frequently used *multivariate Slivnyak-Mecke formula* by a simple induction. Here a point $s \in S^n$ is written as $(s_1, ..., s_n)$.

Theorem 4.17 (multivariate Slivnyak-Mecke formula). *Let ξ be a Poisson process on a Borel space S with intensity measure μ. Let $n \in \mathbb{N}$ and $f : \mathbf{M}(S) \times S^n \to [0, \infty]$ be a measurable function. Then*

$$\mathbb{E} \int f(\xi, s) \xi^{(n)}(ds) = \int \mathbb{E} f\left(\xi + \sum_{i=1}^{n} \delta_{s_i}, s\right) \mu^n(ds). \tag{4.13}$$

4.2.2 Cox processes

Let $(\Omega, \mathcal{A}, \mathbb{P})$ denote a basic probability space. A *uniformly σ-finite* random measure η on a Borel space S is a random measure on S such that there is a partition $P := (B_1, B_2, \dots)$ of S where $\eta(B_i) < \infty$ $\mathbb{P}$-a.e. for each $i \in \mathbb{N}$. Given such a uniformly σ-finite random measure η, a *Cox process* ξ directed by η may be defined as the random measure derived by the following 2-step stochastic experiment. In the first step we realize the a.s. σ-finite η and in the second step construct ξ as a Poisson process with respect to the previously generated η. It is now a simple consequence, that ξ is uniformly σ-finite: by the Poisson property the uniform σ-finiteness of η directly carries over to ξ and one may even use the same partition for ξ as for η. Thus *both* η and ξ may be interpreted as random elements in the Borel space $\mathbf{M}^P(S)$ and hence even the random pair (ξ, η) is essentially a random element in a Borel space. This will be crucial for us in the next theorems since it allows us to condition (ξ, η) on arbitrary random elements, in particular on η itself. We note that a uniformly σ-finite random measure is Cox driven by η if and only if conditional on η it is a Poisson process with intensity measure η and in this case we clearly have $\mathbb{E}[\xi(\cdot)|\eta] = \eta(\cdot)$ and thus trivially $\mathbb{E}\xi(\cdot) = \mathbb{E}\eta(\cdot)$.

We first derive by a simple conditioning procedure the following extension of Mecke's characterization to Cox processes.

Theorem 4.18 (characterization of Cox-processes). *Consider a point process ξ and a uniformly σ-finite random measure η on a Borel space S. Then ξ is a Cox process driven by η if and only if*

$$\mathbb{E} \int f(\xi, \eta, s) \xi(ds) = \mathbb{E} \int f(\xi + \delta_s, \eta, s) \eta(ds), \quad f \in (\mathcal{M}(S)^2 \otimes \mathcal{S})_+. \tag{4.14}$$

Proof. First assume that ξ is Cox driven by η. Then from what has been said above the theorem we may condition the left side of (4.14) on η which gives

$$\mathbb{E} \int f(\xi, \eta, s) \xi(ds) = \mathbb{E} \left[\mathbb{E} \left[\int f(\xi, \eta, s) \xi(ds) \Big| \eta \right] \right].$$

Since conditional on η the point process ξ is Poisson with intensity measure η, Theorem 4.16 gives

$$\mathbb{E} \int f(\xi, \eta, s) \xi(ds) = \mathbb{E} \left[\mathbb{E} \left[\int f(\xi + \delta_s, \eta, s) \eta(ds) \Big| \eta \right] \right]$$

and this clearly reduces to the right side of (4.14). Conversely, if (4.14) holds then in particular

$$\mathbb{E} \int g(\eta) f(\xi, s) \xi(ds) = \mathbb{E} \int g(\eta) f(\xi + \delta_s, s) \eta(ds)$$

for arbitrary $g \in (\mathcal{M}(S))_+$ and $f \in (\mathcal{M}(S) \otimes \mathcal{S})_+$. Thus conditioning on η yields

$$\mathbb{E} \left[g(\eta) \mathbb{E} \left[\int f(\xi, s) \xi(ds) \Big| \eta \right] \right] = \mathbb{E} \left[g(\eta) \mathbb{E} \left[\int f(\xi + \delta_s, s) \eta(ds) \Big| \eta \right] \right],$$

and since all factors are η-measurable and g was arbitrary, it follows that

$$\mathbb{E} \left[\int f(\xi, s) \xi(ds) \Big| \eta \right] = \mathbb{E} \left[\int f(\xi + \delta_s, s) \eta(ds) \Big| \eta \right] \quad \mathbb{P}\text{-a.e.}$$

Theorem 4.16 now implies that a.s. ξ is conditional on η a Poisson process with (conditional) intensity measure η and thus that it is Cox driven by η. □

It is a small step to extend the argument used in the previous proof to the multivariate case.

Theorem 4.19 (multivariate Cox formula). *A Cox process ξ on a Borel space S driven by η satisfies for any $n \in \mathbb{N}$*

$$\mathbb{E} \int f(\xi, \eta, s) \xi^{(n)}(ds) = \mathbb{E} \int f\left(\xi + \sum_{i=1}^{n} \delta_{s_i}, \eta, s \right) \eta^n(ds), \quad f \in (\mathcal{M}(S)^2 \otimes \mathcal{S}^n)_+.$$

(4.15)

Proof. It is enough to condition the left side of (4.15) on η and to apply Theorem 4.17. □

Remark 4.20. We note that Theorem 4.19 yields in particular the relation

$$\mathbb{E}\xi^{(n)} = \mathbb{E}\eta^n,$$

which allows us to choose a σ-finite supporting measure $\nu_n \sim \mathbb{E}\xi^{(n)} = \mathbb{E}\eta^n$ for both. Applying the Palm disintegration (2.22) with this same ν_n on both sides of equation (4.15) yields

$$\mathbb{P}\left((\xi, \eta) \in \cdot \Big\| \xi^{(n)} \right)_s = \mathbb{P}\left(\left(\xi + \sum_{i=1}^{n} \delta_{s_i}, \eta \right) \in \cdot \Big\| \eta^n \right)_s, \quad \nu_n\text{-a.e. } s \in S^n, \quad (4.16)$$

where ν-a.e. means evidently the same as $\mathbb{E}\xi^{(n)}$-a.e. or $\mathbb{E}\eta^n$-a.e. This slightly extends [33, Theorem 4.2 (ii), (iii)] which contains the two statements that one gets from (4.16) by forming the two marginals.

4.2.3   Partially stationary Cox processes

Considering a fixed group action $G \hookrightarrow S$ and assuming the existence of a fixed measurable system O of orbit representatives in S we may formulate a Slivnyak-type result for general G-stationary Cox processes on S by using the cumulative Palm measure of ξ with respect to O. Clearly the characterizing equation (4.14) also applies in this case, but the stationarity allows us to drop, intuitively speaking, all but one of the Palm distributions from each orbit respectively. Even though this is basically known from the transitive case, the non-transitive case seems to be at least technically new (the results of Kallenberg [30] already suggest the following result on an intuitive level since Palm kernels may be chosen to be G-invariant, i.e. no information is lost when dropping all but one Palm measure in every single orbit) and the neat statement using the cumulative Palm measure is of independent interest to us. We recall that Theorem 4.10 establishes the link to Palm pairs and emphasize that it makes the above stated intuitive reduction of Palm kernel members precise. The identity map on O shall be denoted by $\mathbf{b}$, i.e. $\mathbf{b}(b) = b, b \in O$.

Theorem 4.21. (cumulative Palm measure and G-stationary Cox Processes) *Let ξ be a G-stationary point process and η a G-stationary random measure with σ-finite intensity measure both living on a Borel space S. Then ξ is Cox driven by η if and only if*

$$\mathbb{Q}^{\xi}\left((\xi, \eta, \mathbf{b}) \in \cdot\right) = \mathbb{Q}^{\eta}\left((\xi + \delta_{\mathbf{b}}, \eta, \mathbf{b}) \in \cdot\right). \tag{4.17}$$

Proof. Assume first that ξ is Cox driven by η. Then in particular $\mathbb{E}\xi = \mathbb{E}\eta$ is σ-finite. Since by (4.4)

$$\mathbb{Q}^{\xi}\left((\xi, \eta, \mathbf{b}) \in \cdot\right) = \mathbb{E}\iint \mathbf{1}\left\{\left(\xi\left(\theta_g^{-1}\right), \eta\left(\theta_g^{-1}\right), \beta(s)\right) \in \cdot\right\} \kappa_{\beta(s),s}(dg)w(s)\xi(ds)$$

$$= \mathbb{E}\iint \mathbf{1}\left\{\left(g^{-1}\xi, g^{-1}\eta, \beta(s)\right) \in \cdot\right\} \kappa_{\beta(s),s}(dg)w(s)\xi(ds),$$

equation (4.14) implies that $\mathbb{Q}^{\xi}((\xi, \eta, \mathbf{b}) \in \cdot)$ equals

$$\mathbb{E}\iint \mathbf{1}\left\{\left(g^{-1}(\xi + \delta_s), g^{-1}\eta, \beta(s)\right) \in \cdot\right\} \kappa_{\beta(s),s}(dg)w(s)\eta(ds).$$

Noting that $g^{-1}\delta_s(\cdot) = \delta_s(g\cdot) = \delta_{g^{-1}s}(\cdot) = \delta_{\beta(s)}(\cdot)$ for $g \in G_{\beta(s),s}$ and using the G-stationarity of ξ and η gives that the above equals

$$\mathbb{E}\iint \mathbf{1}\left\{\left(\xi(\theta_g^{-1}) + \delta_{\beta(s)}, \eta(\theta_g^{-1}), \beta(s)\right) \in \cdot\right\} \kappa_{\beta(s),s}(dg)w(s)\eta(ds),$$

and this equals the right side of (4.17) by (4.4).

Conversely if (4.17) holds, then

$$\iint f(g\xi(\omega), g\eta(\omega), gb)\lambda(dg)\mathbb{Q}^{\xi}(d(\omega, b))$$

$$= \iint f(g(\xi(\omega) + \delta_b), g\eta(\omega), gb)\lambda(dg)\mathbb{Q}^{\eta}(d(\omega, b))$$

which means

$$\iint f(\xi(\theta_g\omega), \eta(\theta_g\omega), gb)\lambda(dg)\mathbb{Q}^{\xi}(d(\omega, b))$$

$$= \iint f(\xi(\theta_g\omega) + \delta_{gb}, \eta(\theta_g\omega), gb)\lambda(dg)\mathbb{Q}^{\eta}(d(\omega, b)).$$

Applying (4.2) on the left side with respect to ξ and on the right side with respect to η yields (4.14) and thus by Theorem 4.18, ξ is Cox driven by η. $\square$

The special case when ξ is Poisson (i.e. when $\eta = \mu$ is deterministic) deserves a separate formulation as the argument η may be dropped and the right side simplifies a bit:

Corollary 4.22. (cumulative Palm measure and G-stationary Poisson Processes) *Let ξ be a G-stationary point process on the Borel space S with σ-finite intensity measure. Then ξ is a Poisson process if and only if*

$$\mathbb{Q}^{\xi}((\xi, \mathbf{b}) \in \cdot) = \mathbb{Q}^{\mathbb{E}\xi}((\xi + \delta_{\mathbf{b}}, \mathbf{b}) \in \cdot). \tag{4.18}$$

In addition

$$\mathbb{Q}^{\mathbb{E}\xi}((\xi + \delta_{\mathbf{b}}, \mathbf{b}) \in \cdot) = \int \mathbb{P}((\xi + \delta_b, b) \in \cdot)(\mathbb{E}\xi)^*(db). \tag{4.19}$$

Proof. Everything besides (4.19) follows from Theorem 4.21 and to see that last assertion we note that from (4.4) and Fubini's theorem

$$\mathbb{Q}^{\mathbb{E}\xi}((\xi + \delta_{\mathbf{b}}, \mathbf{b}) \in \cdot) = \int \mathbb{P}((\xi(\theta_g^{-1}) + \delta_{\beta(s)}, \beta(s)) \in \cdot)\kappa_{\beta(s),s}(dg)w(s)(\mathbb{E}\xi)(ds).$$

It remains to use G-invariance of $\mathbb{P}$ and to use the decomposition of $\mathbb{E}\xi$ as in (2.7). $\square$

4.3   Palm probability measures

Given a G-stationary random measure on the Borel space S the cumulative Palm measure of ξ with respect to an arbitrary fixed system of orbit representatives O is usually not a finite measure. To see this we note that by (4.4) it carries total mass

$$\mathbb{Q}(\Omega \times S) = \mathbb{Q}(\Omega \times O) = \mathbb{E} \int w(s)\xi(ds) = (\mathbb{E}\xi)^*(O) \tag{4.20}$$

where w is such that $\mu_b w = 1, b \in O$ (and besides that arbitrary). This is infinite for a large class of operations $G \hookrightarrow S$. Still it will be possible to extract meaningful and interesting probability measures from it by restricting it to certain subsets. This will be the content of Subsection 4.3.1.

4.3.1   Cumulative Palm probability measures

We note that for any jointly G-invariant measurable subset $I \subset \Omega \times S$ the restriction $\mathbb{Q}(\cdot \cap I)$ of $\mathbb{Q}$ to I is by (4.4) equal to the cumulative Palm measure of the random measure

$$\xi_I(\omega, \cdot) := \int \mathbf{1}\{s \in \cdot\}\mathbf{1}_I(\omega, s)\xi(\omega, ds)$$

and carries again by (4.4) the total mass

$$\mathbb{Q}(I) = \mathbb{E} \int \mathbf{1}_I(\theta_e, s)w(s)\xi(ds).$$

As we will see in plenty of examples later, I may often be chosen such that $0 < \mathbb{Q}(I) < \infty$. In any such case we may define the following probability measure.

Definition 4.23 (cumulative Palm probability measure). Given a G-stationary random measure ξ on the Borel space S and a jointly G-invariant measurable set $I \subset \Omega \times S$ with $0 < \mathbb{Q}(I) < \infty$ we define the *I-averaged Palm probability measure* $\mathbb{P}^I$ on $\Omega \times S$ via

$$\mathbb{P}^I(\cdot) := \frac{1}{\mathbb{Q}(I)} \mathbb{Q}(\cdot \cap I). \tag{4.21}$$

It is clearly concentrated on $\tilde{I} := (\Omega \times O) \cap I$ and thus

$$\left(\tilde{I}, (\mathcal{A} \otimes \mathcal{S})|\tilde{I}, \mathbb{P}^I \right) \tag{4.22}$$

may be considered as an underlying basic probability space, where the expression in the middle denotes the trace of $\mathcal{A} \otimes \mathcal{S}$ on $\tilde{I}$. Since $\mathbb{Q}(\cdot \cap I)$ is the cumulative Palm measure of ξ_I we may interpret $\mathbb{P}^I$ as follows in many special cases. Namely, if $\mathbb{E}\xi_I$ is σ-finite and Ω is Borel, Theorem 4.10 yields the disintegration

$$\mathbb{P}^I = \frac{1}{\mathbb{Q}(I)} \iint \mathbf{1}\{(\omega, b) \in \cdot\} \mathbb{P}\left(d\omega \| \xi_I\right)_b (\mathbb{E}\xi_I)^*(db).$$

It follows that the finite and non-zero number $\mathbb{Q}(I)$ is given by

$$\mathbb{Q}(I) = (\mathbb{E}\xi_I)^*(O), \tag{4.23}$$

which means

$$\mathbb{P}^I(\cdot) = \iint \mathbf{1}\{(\omega, b) \in \cdot\} \mathbb{P}\left(d\omega \| \xi_I\right)_b (d\omega) \frac{(\mathbb{E}\xi_I)^*(db)}{(\mathbb{E}\xi_I)^*(O)}. \tag{4.24}$$

Thus $\mathbb{P}^I$ governs a 2-step stochastic experiment. First a random orbit (representative) $\mathbf{b}$ is chosen according to

$$\frac{(\mathbb{E}\xi_I)^*(\cdot)}{(\mathbb{E}\xi_I)^*(O)}$$

and then the configuration ω is picked according to the Palm probability measure

$$P_{\mathbf{b}}^{\xi|I}(\cdot) = \mathbb{P}\left(\cdot \| \xi_I \right)_{\mathbf{b}}$$

at $\mathbf{b}$ with respect to ξ_I. $\mathbb{P}^I$ is the joint law of these two random elements. Even the case without stationarity ($G = \{e\}$) is interesting. Here $O = S$, $(\mathbb{E}\xi_I)^* = \mathbb{E}\xi_I$, $w = 1$ and the joint G-invariance requirement upon $I \subset \Omega \times S$ represents a condition that is always fulfilled. Hence $0 < \mathbb{Q}(I) < \infty$ holds if and only if $0 < \mathbb{E}\xi_I(S) < \infty$ and in this case $\mathbf{b}$ is chosen according to

$$\frac{(\mathbb{E}\xi_I)(\cdot)}{(\mathbb{E}\xi_I)(S)}.$$

If in this case $I = \Omega \times A$ where $A \in \mathcal{S}$ is arbitrary, then $\mathbf{b}$ is chosen according to

$$\frac{(\mathbb{E}\xi)(\cdot \cap A)}{(\mathbb{E}\xi)(A)}.$$

If η is a random element in a Borel space T such that (η, ξ) is jointly G-stationary then the distribution of (η, ξ) under $\mathbb{P}^I$ gets a similar interpretation, where it suffices now to require $\mathbb{E}\xi_I$ to be σ-finite (Ω need not be Borel). Here Theorem 4.12 yields with similar steps as above

$$\mathbb{P}^I\left((\eta, \xi) \in \cdot\right) = \iint \mathbf{1}\{(t, b) \in \cdot\}\mathbb{P}\left(\eta \in dt || \xi_I\right)_b \frac{(\mathbb{E}\xi_I)^*(db)}{(\mathbb{E}\xi_I)^*(O)}. \qquad (4.25)$$

Example 4.24 ($I = \Omega \times A$ where A invariant). If the jointly G-invariant set I is of the special form $I = \Omega \times A$ where $A \in \mathcal{S}$ is G-invariant then $(\mathbb{E}\xi_I)^* = (\mathbb{E}\xi)^*|A$ and (4.23) simplifies to

$$\mathbb{Q}(I) = (\mathbb{E}\xi)^*(A).$$

Hence, if ξ and A are such that the *finite height condition* $0 < (\mathbb{E}\xi)^*(A) < \infty$ is satisfied then $\mathbb{P}^I$ is defined and may be written as

$$\mathbb{P}^A(\cdot) := \mathbb{P}^I(\cdot) = \frac{1}{(\mathbb{E}\xi)^*(A)} \int \mathbf{1}\{(\omega, b) \in \cdot\}\mathbf{1}_A(b)\mathbb{Q}(d(\omega, b)). \qquad (4.26)$$

The equations (4.24) and (4.25) reduce to

$$\mathbb{P}^I(\cdot) = \iint \mathbf{1}\{(\omega, b) \in \cdot\}\mathbb{P}\left(d\omega || \xi\right)_b \frac{\mathbf{1}_A(b)(\mathbb{E}\xi)^*(db)}{(\mathbb{E}\xi)^*(A)},$$

and

$$\mathbb{P}^I\left((\eta, \xi) \in \cdot\right) = \iint \mathbf{1}\{(t, b) \in \cdot\}\mathbb{P}\left(\eta \in dt || \xi\right)_b \frac{\mathbf{1}_A(b)(\mathbb{E}\xi)^*(db)}{(\mathbb{E}\xi)^*(A)}.$$

4.3.2 Conditional cumulative Palm measures

The defining equation (4.21) already suggests a close link between cumulative Palm probability measures and ordinary conditional probabilities.

Lemma 4.25 (conditioning on jointly invariant subsets). *Suppose $I_1, I_2 \in \mathcal{A} \otimes \mathcal{S}$ are both jointly G-invariant such that $I_1 \subset I_2$ and $0 < \mathbb{Q}(I_1), \mathbb{Q}(I_2) < \infty$. Then for any G-stationary random measure ξ on the Borel space S*

$$\mathbb{P}^{I_1}(\cdot) = \mathbb{P}^{I_2}(\cdot | I_1).$$

Proof. By definition and $I_1 \subset I_2$ we have

$$\mathbb{P}^{I_1}(\cdot) = \frac{\mathbb{Q}(\cdot \cap I_1)}{\mathbb{Q}(I_1)} = \frac{\mathbb{Q}(\cdot \cap I_1 \cap I_2)}{\mathbb{Q}(I_1 \cap I_2)} = \frac{\mathbb{P}^{I_2}(\cdot \cap I_1)}{\mathbb{P}^{I_2}(I_1)} = \mathbb{P}^{I_2}(\cdot | I_1). \qquad \square$$

Example 4.26 (partially stationary tessellations). We consider the operation $L \hookrightarrow \mathbb{R}^d$ of a fixed linear subspace L of dimension $0 \le u \le d$ of $\mathbb{R}^d$ on $\mathbb{R}^d$ via translation together with an L-stationary simple point process ξ in $\mathbb{R}^d$. This point process induces a random *Voronoi tessellation* of $\mathbb{R}^d$ defined as the collection of *Voronoi cells*

$$C(\omega, s) := \{x \in \mathbb{R}^d : ||x - s|| \leq ||x - y||, y \in \xi(\omega)\}, \quad s \in \xi(\omega), \omega \in \Omega,$$

and $C(\omega, s) := \emptyset$ if $s \notin \xi(\omega)$. Clearly

$$C(\theta_l \omega, l + s) = l + C(\omega, s), \quad l \in L, s \in S, \omega \in \Omega,$$

and we naturally interpret $s \in \xi(\omega)$ as the *center* of the cell $C(\omega, s)$ in configuration ω. Fixing an L-invariant subset A of $\mathbb{R}^d$ we may consider the following three different jointly L-invariant subsets I_1, I_2 and I_3 of $\Omega \times S$ defined via

$$I_1 := \{(\omega, s) : \emptyset \neq C(\omega, s) \subset A\},$$

$$I_2 := \{(\omega, s) : \emptyset \neq C(\omega, s), s \in A\},$$

and

$$I_3 := \{(\omega, s) : C(\omega, s) \cap A \neq \emptyset\}.$$

Clearly $I_1 \subset I_2 \subset I_3$ and Lemma 4.25 applies. This gives

$$\mathbb{P}^{I_1}(\cdot) = \mathbb{P}^{I_2}(\cdot|I_1) = \mathbb{P}^{I_3}(\cdot|I_1) \quad \text{and} \quad \mathbb{P}^{I_2}(\cdot) = \mathbb{P}^{I_3}(\cdot|I_2).$$

Here a random polyhedral set Z

(1) with distribution $\mathbb{P}^{I_1}(C(\theta_e, \mathbf{b}) \in \cdot)$ may be interpreted as the *typical cell of the Voronoi tessellation under all cells contained in A,*

(2) with distribution $\mathbb{P}^{I_2}(C(\theta_e, \mathbf{b}) \in \cdot)$ may be interpreted as the *typical cell of the Voronoi tessellation under all cells with center in A,*

(3) with distribution $\mathbb{P}^{I_3}(C(\theta_e, \mathbf{b}) \in \cdot)$ may be interpreted as the *typical cell of the Voronoi tessellation under all cells intersecting A.*

Clearly we did not use any special property of *Voronoi* tessellations here and it should be clear how to extend these notions to general L-stationary tessellations. We shall inspect these notions in more detail for Cox-Voronoi mosaics in Section 7.1.

4.4 Properties of cumulative Palm measures

This sections contains first two results from [21] that have been also proved independently by Kallenberg [31]: the *Transport Theorem* 4.27, generalizing [39, Theorem 3.6], [37, Theorem 3.15] and [38, Theorem 4.1] to non-transitive underlying group actions, and the *Characterization Theorem* 4.33, generalizing Mecke's [46] famous characterization of the classical Palm measure to a characterization of our cumulative Palm measure for possibly non-transitive group actions. Second, we extend [38, Theorem 5.1] of Last in Theorem 4.31 which characterizes balancing G-invariant transports between G-stationary random measures ξ and η in terms of a transport relation between their respective cumulative Palm measures.

4.4.1   The Transport Theorem

As before we consider a lcsc group G operating measurably on Ω and properly on the Borel spaces $(S, \mathcal{S})$ and $(T, \mathcal{T})$. For the sake of generality we shall allow $\mathbb{P}$ to be a σ-finite measure on $(\Omega, \mathcal{A})$. Our aim is to derive a fundamental transport property of Palm measures. In the special case where $G = S = T$ is an Abelian group the result boils down to Theorem 3.6 in [39]. Other special cases will be discussed below. Recall the function Δ^* defined as in (3.9). We consider two G-stationary random measures ξ on S and η on T, together with invariant kernels γ from $\Omega \times S$ to T and δ from $\Omega \times T$ to S. Here invariance is to be interpreted with respect to the diagonal operation whenever product spaces come into play, i.e. we require e.g. γ to satisfy

$$\gamma(\theta_g \omega, gs, B) = \gamma(\omega, s, g^{-1}B), \quad g \in G, s \in S, \omega \in \Omega, B \in \mathcal{S}. \tag{4.27}$$

Consider the balance equation

$$\iint \mathbf{1}\{(s, t) \in \cdot\} \gamma(\omega, s, dt) \xi(\omega, ds) = \iint \mathbf{1}\{(s, t) \in \cdot\} \delta(\omega, t, ds) \eta(\omega, dt), \quad \mathbb{P}\text{-a.e. } \omega \in \Omega.$$

Intuitively the applications of the random kernel γ to the random measure ξ on S should be interpreted as lifting ξ to a random measure Γ on $S \times T$ where $\Gamma(\cdot \times T)$ is a possibly resized version of ξ while $\Gamma(S \times \cdot)$ is a possibly resized version of η. Further, the application of δ to η must lift to the very same Γ for $\mathbb{P}$-a.e. $\omega \in \Omega$. Given fixed systems of orbit representatives $O_S \subset S$ and $O_T \subset T$ we denote the corresponding choice functions by β^S and β^T. We will skip the upper indices whenever there is no risk of confusion. The following theorem has been proved independently by Gentner and Last [21] and Kallenberg [31].

Theorem 4.27 (Transport theorem). *Consider two invariant random measures ξ and η on S and T respectively, let γ and δ be invariant kernels from $\Omega \times S$ to T and from $\Omega \times T$ to S respectively satisfying*

$$\iint \mathbf{1}\{(s, t) \in \cdot\} \gamma(\omega, s, dt) \xi(\omega, ds) = \iint \mathbf{1}\{(s, t) \in \cdot\} \delta(\omega, t, ds) \eta(\omega, dt) \tag{4.28}$$

for $\mathbb{P}$-a.e. $\omega \in \Omega$, and let $\mathbb{Q}^\xi$ and $\mathbb{Q}^\eta$ be the cumulative Palm measures of ξ and η respectively, with respect to fixed systems of orbit representatives O_S and O_T. Then we have for any measurable function $f \in (\mathcal{A} \otimes \mathcal{G} \otimes \mathcal{S} \otimes \mathcal{T})_+$ that

$$\iiint f(\omega, g, b, \beta^T(t)) \kappa_{\beta(t),t}(dg) \gamma(\omega, b, dt) \mathbb{Q}^\xi(d(\omega, b))$$

$$= \iiint f(\theta_g^{-1}\omega, g^{-1}, \beta^S(s), b) \Delta^*(s) \kappa_{\beta(s),s}(dg) \delta(\omega, b, ds) \mathbb{Q}^\eta(d(\omega, b)). \tag{4.29}$$

Proof. Let $w : S \to (0, \infty)$ be as in (2.6). Then for any $b \in O_S$ and $g \in G$

$$\int w(g^{-1}hb) \lambda(dh) = 1.$$

Take $f \in (\mathcal{A} \otimes \mathcal{G} \otimes \mathcal{S} \otimes \mathcal{T})_+$ and denote the left-hand side of (4.29) by I. By Fubini's theorem,

$$I = \iiiint f(\omega, g, b, \beta(t)) w(g^{-1}hb) \kappa_{\beta(t),t}(dg) \gamma(\omega, b, dt) \lambda(dh) \mathbb{Q}^\xi(d(\omega, b)).$$

Applying the refined Campbell theorem (4.11) gives that I equals

$$\mathbb{E} \iiiint f(\theta_h^{-1}, g, \beta(s), \beta(t)) w(g^{-1} h \beta(s)) \kappa_{\beta(t),t}(dg) \gamma(\theta_h^{-1}, \beta(s), dt) \kappa_{\beta(s),s}(dh) \xi(ds)$$

$$= \mathbb{E} \iiiint f(\theta_h^{-1}, g, \beta(s), \beta(t)) w(g^{-1} s) \kappa_{\beta(t), h^{-1}t}(dg) \gamma(h\beta(s), dt) \kappa_{\beta(s),s}(dh) \xi(ds)$$

$$= \mathbb{E} \iiiint f(\theta_h^{-1}, g, \beta(s), \beta(t)) w(g^{-1} s) \kappa_{\beta(t), h^{-1}t}(dg) \kappa_{\beta(s),s}(dh) \gamma(s, dt) \xi(ds),$$

where we used invariance (4.27) of γ and β and the fact that $\kappa_{\beta(s),s}$ is concentrated on $G_{\beta(s),s}$ (see Theorem 3.1 (ii)). By Theorem 3.1 (i) and (4.28)

$$I = \mathbb{E} \iiiint f(\theta_h^{-1}, h^{-1}g, \beta(s), \beta(t)) w(g^{-1} hs) \kappa_{\beta(s),s}(dh) \delta(t, ds) \kappa_{\beta(t),t}(dg) \eta(dt).$$

Using invariance of δ and κ, we obtain that I equals

$$\mathbb{E} \iiiint f(\theta_h^{-1} \circ \theta_g^{-1}, h^{-1}, \beta(s), \beta(t)) w(hgs) \kappa_{\beta(s),s}(dh) \delta(\theta_g^{-1}, \beta(t), ds) \kappa_{\beta(t),t}(dg) \eta(dt),$$

where we have used that $\theta_{gh}^{-1} = \theta_h^{-1} \circ \theta_g^{-1}$ and that $g^{-1}t = \beta(t)$ for t, g as in the above integral. At this stage we can use the refined Campbell Formula (4.11) for η to obtain that I equals

$$\iiiint f(\theta_h^{-1}, h^{-1}, b, \beta(t)) w(hgs) \kappa_{\beta(s),s}(dh) \delta(b, ds) \lambda(dg) \mathbb{Q}^\eta(d(\omega, b)).$$

Now take $h \in G$ and $s \in S$ with $h\beta(s) = s$. Then

$$\int w(hgs) \lambda(dg) = \int w(gh\beta(s)) \lambda(dg) = \Delta(h^{-1}) = \Delta^*(s),$$

where we used Lemma 3.12. Hence we obain from Fubini's theorem that I equals the right-hand side of (4.29). $\qquad \square$

An immediate consequence of Theorem 4.27 is the following *exchange formula* for cumulative Palm measures. A first version of this fundamental and very useful formula was obtained by Neveu (see e.g. [56]) for ordinary Palm measures of random measures on Abelian groups.

Corollary 4.28 (exchange formula). *Let ξ and η be G-invariant random measures on S and T respectively. Then for any $f \in (\mathcal{A} \otimes \mathcal{G} \otimes \mathcal{S} \otimes \mathcal{T})_+$*

$$\iiint f(\omega, g, b, \beta(t)) \kappa_{\beta(t),t}(dg) \eta(\omega, dt) \mathbb{Q}^\xi(d(\omega, b))$$

$$= \iiint f(\theta_g^{-1} \omega, g^{-1}, \beta(s), b) \Delta^*(s) \kappa_{\beta(s),s}(dg) \xi(\omega, ds) \mathbb{Q}^\eta(d(\omega, b)). \tag{4.30}$$

Proof. Specialize $\gamma(\omega, s, dt) := \eta(\omega, dt)$ and $\delta(\omega, t, ds) := \xi(\omega, ds)$. $\qquad \square$

Another consequence is the following formula that arises when using integrands of a special form in (4.28).

Corollary 4.29 (special integrands). *Under the hypothesis of Theorem 4.27 we have for any $f \in (\mathcal{A} \otimes \mathcal{S} \otimes \mathcal{T})_+$*

$$\iint f(\omega, b, t) \gamma(\omega, b, dt) \mathbb{Q}^\xi(d(\omega, b))$$

$$= \iiint f(\theta_g^{-1}\omega, \beta^S(s), g^{-1}b) \Delta^*(s) \kappa_{\beta(s),s}(dg) \delta(\omega, b, ds) \mathbb{Q}^\eta(d(\omega, b)). \tag{4.31}$$

Proof. For arbitrary $\tilde{f} \in (\mathcal{A} \otimes \mathcal{S} \otimes \mathcal{T})_+$ we may take $f(\omega, g, s, t) := \tilde{f}(\omega, s, gt)$ in (4.28). $\qquad \square$

4.4.2   Transport properties

We consider Borel spaces S and T and an lcsc group G operating on each of them properly. According to Lemma 2.21 a σ-finite kernel τ from $\Omega \times S$ to T may be used to transform a random measure ξ on S (i.e. intuitively speaking a random 'mass distribution' on S) into a random measure on T by forming

$$\int \tau(\omega, s, \cdot) \xi(\omega, ds).$$

This underlying intuition for τ of resizing and redistributing one random mass configuration into another one in a random way is the reason why such kernels τ are sometimes referred to as *weighted transport kernels* ([37, 38, 39]). Fixing another random measure η on T we say, following Last [37, 38] and Last and Thorisson [39], that τ is (ξ, η)-*balancing* if

$$\int \tau(\omega, s, \cdot) \xi(ds) = \eta(\omega, \cdot), \quad \omega \in \Omega,$$

and $\mathbb{P}$-*a.e.* (ξ, η)-*balancing* if the above holds only for $\mathbb{P}$-a.e. ω.

In view of the Transport Theorem 4.27 it is of interest to know whether or not for a (ξ, η)-balancing G-invariant transport τ a G-invariant *inverse transport kernel* exists. This is a kernel τ^*, that for given ξ, η and $\gamma := \tau$ satisfies (4.28) when putting $\delta := \tau^*$. The answer is positive and may be derived by a straightforward adaption of arguments found in Last [37, 38].

Lemma 4.30 (existence of inverse transports). *Let ξ and η denote G-stationary random measures on the Borel spaces S and T respectively and let τ denote a (ξ, η)-balancing weighted transport kernel. Then there is a G-invariant Markovian transport kernel τ^* from $\Omega \times T$ to S such that (4.28) holds $\mathbb{P}$-a.s., i.e. a stochastic inverse transport kernel.*

Proof. The measure M on $\Omega \times S \times T$ defined by

$$M := \iiint \mathbf{1}\{(\omega, s, t) \in \cdot\} \tau(\omega, s, dt) \xi(\omega, ds) \mathbb{P}(d\omega)$$

is σ-finite, since it is the Campbell measure of the random measure $\xi \otimes \tau$ (see Lemma 2.21 (ii) and Lemma 2.19 (i)) and jointly G-invariant, as is easy to check. Moreover

$$\int \mathbf{1}\{(\omega, t) \in \cdot\} M(d(\omega, s, t)) = C_\eta(\cdot)$$

since τ is $\mathbb{P}$-a.e. (ξ, η)-balancing, which is again σ-finite by Lemma 2.19 (i) and jointly G-invariant. Thus Theorem 2.16 yields a stochastic G-invariant kernel τ^* from $\Omega \times T$ to S with

$$M = C_\eta \otimes \tau^*.$$

This means

$$\iiint \mathbf{1}\{(\omega, s, t) \in \cdot\} \tau(\omega, s, dt) \xi(\omega, ds) \mathbb{P}(d\omega)$$
$$= \iiint \mathbf{1}\{(\omega, s, t) \in \cdot\} \tau^*(\omega, t, ds) \eta(\omega, dt) \mathbb{P}(d\omega),$$

which implies the assertion since $\mathcal{S} \otimes \mathcal{T}$ is countably generated. □

The next theorem clarifies how the cumulative Palm measure transforms under such transport kernels under invariance assumptions and represents a modest extension of results of Last in [38] and [37] from the case of random measures on one homogeneous space to the case of possibly non-transitive group operations on two possibly different spaces.

Theorem 4.31 (cumulative Palm measure and balancing transports). *Let G operate properly on the Borel spaces S and T and consider G-stationary random measures ξ and η on S and T respectively. Then an invariant weighted transport kernel τ from $\Omega \times S$ to T is $\mathbb{P}$-a.e. (ξ, η)-balancing if and only if*

$$\iiint f(\theta_g^{-1}\omega, \beta^T(t)) \Delta^*(t) \kappa_{\beta(t),t}(dg) \tau(\omega, b, dt) \mathbb{Q}^\xi(d(\omega, b)) = \mathbb{Q}^\eta f \qquad (4.32)$$

for any measurable $f : \Omega \times O_T \to [0, \infty)$.

Proof. Assume first that τ is $\mathbb{P}$-a.e. (ξ, η)-balancing. Then Lemma 4.30 yields a stochastic invariant kernel τ^* from $\Omega \times T$ to S satisfying

$$\iint \mathbf{1}\{(s, t) \in \cdot\} \tau(\omega, s, dt) \xi(\omega, ds) = \iint \mathbf{1}\{(s, t) \in \cdot\} \tau^*(\omega, t, ds) \eta(\omega, dt).$$

The Transport Theorem 4.27 gives for any $f \in (\mathcal{A} \otimes \mathcal{G} \otimes \mathcal{S} \otimes \mathcal{T})_+$ that

$$\iiint f(\theta_g^{-1}\omega, g^{-1}, b, \beta^T(t)) \Delta^*(t) \kappa_{\beta(t),t}(dg) \tau(\omega, b, dt) \mathbb{Q}^\xi(d(\omega, b))$$
$$= \iiint f(\omega, g, \beta^S(s), b) \kappa_{\beta(s),s}(dg) \tau^*(\omega, b, ds) \mathbb{Q}^\eta(d(\omega, b)).$$

Dropping the second and third argument of f yields (4.32).

Conversely if (4.32) holds for all measurable functions $f : \Omega \times O_T \to [0, \infty)$ then, in view of Lemma 2.19, it is enough to prove

$$\mathbb{E} \iint f(\theta_e, t) \tau(s, dt) \xi(ds) = \mathbb{E} \int f(\theta_e, t) \eta(dt) \qquad (4.33)$$

for all measurable functions $f \in (\mathcal{A} \otimes \mathcal{T})_+$. Starting with the right-hand side we have first by (4.2)

$$\mathbb{E} \int f(\theta_e, t) \eta(dt) = \iint f(\theta_g \omega, gb) \lambda(dg) \mathbb{Q}^\eta(d(\omega, b)).$$

By (4.32) the right side may be written as

$$\iiiint f(\theta_g \theta_h^{-1} \omega, g\beta^T(t))\lambda(dg)\Delta^*(t)\kappa_{\beta(t),t}(dh)\tau(\omega, b, dt)\mathbb{Q}^\xi(d(\omega, b)),$$

where by Lemma 3.12 the term $\Delta^*(t)$ may be replaced by $\Delta(h^{-1})$. Using property (2.1) of the modular function this turns into

$$\iiiint f(\theta_g \omega, gh\beta^T(t))\lambda(dg)\kappa_{\beta(t),t}(dh)\tau(\omega, b, dt)\mathbb{Q}^\xi(d(\omega, b)),$$

which, using the properties of κ, Fubini and invariance of τ, reduces to

$$\iiint f(\theta_g \omega, t)\tau(\theta_g \omega, gb, dt)\lambda(dg)\mathbb{Q}^\xi(d(\omega, b)),$$

and thus to the left-hand side of (4.33) by using (4.2) again. $\qquad\qquad\square$

4.4.3   Characterization

In this subsection we ask which measures on $\Omega \times S$ actually are cumulative Palm measures. The answer will substantially extend Mecke's famous characterization of Palm measures and has been derived (essentially) in Gentner and Last [21] in a different version using Palm pairs and also independently in Kallenberg [31].

We start with the special case where no stationarity or invariance assumptions are made. In view of Example 4.6 the question reads in this case: Which measures on $\Omega \times S$ can be Campbell measures of a given fixed random measure ξ? Note that in contrast to previous sections we do not fix an underlying measure $\mathbb{P}$ on $(\Omega, \mathcal{A})$ here. The answer is given by the following general proposition. Note that we neither have to require a Borel structure on S nor properness or other regularity assumptions on the operation $G \hookrightarrow S$ in part (ii).

Proposition 4.32. (characterization of Campbell measures)

  (i) *A measure C on $\Omega \times S$ is the Campbell measure of a random measure ξ with respect to some underlying σ-finite measure on $(\Omega, \mathcal{A})$ iff C is σ-finite, $C(\{\xi = 0\} \times S) = 0$ and*

$$\iint f(\omega, s, t)\xi(\omega, dt)C(d(\omega, s)) = \iint f(\omega, t, s)\xi(\omega, dt)C(d(\omega, s)) \qquad (4.34)$$

for any $f \in (\mathcal{A} \otimes \mathcal{S} \otimes \mathcal{S})_+$.

  (ii) *If ξ is G-stationary and C is jointly G-invariant then the same characterization holds and in addition the corresponding underlying measure $\mathbb{P}$ on Ω may be chosen G-invariant.*

Proof. (i) First, assume that C actually is the Campbell measure of ξ with respect to some σ-finite measure $\mathbb{P}$ on Ω. As we have seen earlier, every Campbell measure is σ-finite. Further

$$\int \mathbf{1}\{\xi(\omega) = 0\}C(d(\omega, s)) = \mathbb{E}\int \mathbf{1}\{\xi = 0\}\xi(ds) = 0,$$

which yields the second property. To see (4.34) note that

$$\iint f(\omega,s,t)\xi(\omega,dt)C(d(\omega,s)) = \iiint f(\omega,s,t)\xi(\omega,dt)\xi(\omega,ds)\mathbb{P}(d\omega).$$

Hence it is enough to invoke Fubini's Theorem and to interchange the roles of s and t.

To prove the converse assume the three conditions on C. By σ-finiteness we may choose a measurable function $g' > 0$ on $\Omega \times S$ such that $Cg' < \infty$. Since ξ is measurably σ-finite, we may choose by Lemma 2.3 (a) a function $\tilde{g} > 0$ on $\Omega \times S$ with $0 < \xi(\tilde{g}) < \infty$ on $\{\xi \neq 0\}$. Now set $g := g' \wedge \tilde{g}$ and $h(\omega,s) := g(\omega,s)/(\xi g)(\omega)$, where $h(\omega,s) := 0$ if $\xi(g) = 0$. Define the measure $\mathbb{P}$ by

$$\mathbb{P}(A) := \iint \mathbf{1}_A(\omega)h(\omega,s)C(d(\omega,s)), \quad A \in \mathcal{A}. \tag{4.35}$$

By the second assumption we have $\mathbb{P}(\xi = 0) = 0$. Note that $\omega \mapsto \xi(g)(\omega)$ is finite and positive on $\{\xi \neq 0\}$. Furthermore,

$$\mathbb{E}\xi(g)\mathbf{1}\{\xi \neq 0\} = \iint (\xi g)(\omega)\frac{g(\omega,s)}{(\xi g)(\omega)}\mathbf{1}\{\xi \neq 0\}C(d(\omega,s)) \leq \iint g'(\omega,s)C(d(\omega,s)) < \infty.$$

As by the second assumption on C we have $\mathbb{P}(\xi = 0) = 0$ it follows that $\mathbb{P}$ is σ-finite. It remains to prove that indeed C is the Campbell measure of ξ with respect to this $\mathbb{P}$. We have for $f \in (\mathcal{A} \otimes \mathcal{S})_+$ by definition of $\mathbb{P}$

$$\mathbb{E}\int f(\theta_e,t)\xi(dt) = \iint f(\omega,t)h(\omega,s)\xi(\omega,dt)C(d(\omega,s))$$
$$= \iint f(\omega,s)h(\omega,t)\xi(\omega,dt)C(d(\omega,s)),$$

where we have used (4.34) to get the second identity. Since $C(\{\xi = 0\} \times S) = 0$ this may be written as

$$\iint \mathbf{1}\{\xi \neq 0\}f(\omega,s)h(\omega,t)\xi(\omega,dt)C(d(\omega,s))$$

and since $\xi(h) = 1$ on $\{\xi \neq 0\}$ by definition of h this reduces to Cf (again using the assumption $C(\{\xi = 0\} \times S) = 0$).

(ii) It is enough to show that the measure $\mathbb{P}$ defined in (4.35) is invariant for jointly G-invariant C and G-stationary ξ. Take $f \in \mathcal{A}_+$ and $g \in G$. By the joint G-invariance of C

$$\mathbb{E}f \circ \theta_g = \iint f(\theta_g\omega)h(\omega,s)C(d(\omega,s))$$
$$= \iint f(\omega)h(\theta_g^{-1}\omega, g^{-1}s)C(d(\omega,s)).$$

Since C is the Campbell measure of the G-stationary ξ with respect to $\mathbb{P}$ we may proceed:

$$\mathbb{E}f \circ \theta_g = \mathbb{E}\int f(\theta_e)h(\theta_g^{-1}, g^{-1}s)\xi(ds)$$
$$= \iint f(\omega)h(\theta_g^{-1}\omega, s)\xi(\theta_g^{-1}\omega, ds)\mathbb{P}(d\omega) = \mathbb{E}f,$$

where we have used in the last step that $\int h(\theta_g^{-1}\omega, s)\xi(\theta_g^{-1}\omega, ds) = 1$ for $\mathbb{P}$-a.e. ω since $\{\xi \neq 0\}$ is G-invariant and has a complement of $\mathbb{P}$-measure 0. $\qquad\square$

Having characterized Campbell measures which are the cumulative Palm measures of random measures ξ with respect to the trivial operation $\{e\} \hookrightarrow S$, we may proceed now with the general case, independently established by Gentner and Last [21] and Kallenberg [31]. In the transitive special case of Example 4.5 the result has been derived in [60] and [38]. Note that in order to ensure the existence of the cumulative Palm measure we have to require S to be Borel and $G \hookrightarrow S$ to be proper now (in contrast to the previous theorem).

Theorem 4.33 (characterization of cumulative Palm measures). *Consider   the proper operation $G \hookrightarrow S$ where S is Borel together with a fixed measurable system $O = \beta(S)$ of orbit representatives. Let ξ be a G-stationary random measure on S and $\mathbb{Q}$ a measure on $\Omega \times S$. Then $\mathbb{Q}$ is the cumulative Palm measure of ξ with respect to O and some invariant σ-finite measure on $(\Omega, \mathcal{A})$ iff $\mathbb{Q}$ satisfies the stabilizer condition (4.3), is σ-finite, satisfies $\mathbb{Q}(\{\xi = 0\} \times S) = 0$ and for any $f \in (\mathcal{A} \otimes \mathcal{S} \otimes \mathcal{S})_+$*

$$\iiint f(\theta_g^{-1}\omega, \beta(s), g^{-1}b)\Delta^*(s)\kappa_{\beta(s),s}(dg)\xi(\omega, ds)\mathbb{Q}(d(\omega, b))$$
$$= \iint f(\omega, b, s)\xi(\omega, ds)\mathbb{Q}(d(\omega, b)). \quad (4.36)$$

Proof. If $\mathbb{Q}$ is the cumulative Palm measure of ξ with respect to some $O = \beta(S)$ then (4.3) is fulfilled by definition and σ-finiteness of $\mathbb{Q}$ has been proved in Theorem 4.1.1. Further the third property follows from (4.4) and the G-invariance of $\{\xi = 0\}$ since

$$\mathbb{Q}(\{\xi = 0\} \times S) = \mathbb{E} \iint \mathbf{1}\{\xi(\theta_g^{-1}) = 0\}\kappa_{\beta(s),s}(dg)w(s)\xi(ds) = \mathbb{E} \int \mathbf{1}\{\xi = 0\}w(s)\xi(ds) = 0.$$

Equation (4.36) is the special case $T := S$ and $\gamma := \delta := \eta := \xi$ of (4.31).

Conversely assume the regularity conditions and (4.36). We need to verify (4.2) in order to prove that $\mathbb{Q}$ is indeed the cumulative Palm measure of ξ with respect to O and some $\mathbb{P}$. In order to show the existence of a σ-finite $\mathbb{P}$ such that (4.2) holds we may use Proposition 4.32: Consider the measure

$$C := \iint \mathbf{1}\{(\theta_g\omega, gb) \in \cdot\}\lambda(dg)\mathbb{Q}(d(\omega, b))$$

which is evidently σ-finite by σ-finiteness of λ and $\mathbb{Q}$. It also satisfies

$$C(\{\xi = 0\} \times S) = \iint \mathbf{1}\{\xi(\theta_g) = 0\}\lambda(dg)\mathbb{Q}(d(\omega, b)) = \iint \mathbf{1}\{\xi = 0\}\mathbb{Q}(d(\omega, b))\lambda(dg) = 0$$

by the second property of $\mathbb{Q}$. Hence it remains to show that C satisfies (4.34) (then (4.2) follows for some σ-finite $\mathbb{P}$ by Proposition 4.32). We have by G-stationarity of ξ and by definition of C

$$\iint f(\omega, s, t)\xi(\omega, dt)C(d(\omega, s)) = \iiint f(\theta_g\omega, gb, t)\xi(\theta_g\omega, dt)\lambda(dg)\mathbb{Q}(d(\omega, b))$$
$$= \iiint f(\theta_g\omega, gb, gt)\xi(\omega, dt)\lambda(dg)\mathbb{Q}(d(\omega, b)).$$

Using the stochastic kernel κ this last expression can be written as (replacing g by h for better comparability later)

$$\iiiint f(\theta_h\omega, hb, hg\beta(t))\kappa_{\beta(t),t}(dg)\xi(\omega, dt)\lambda(dh)\mathbb{Q}(d(\omega, b)),$$

and this equals

$$\iiiint f(\theta_{hg^{-1}}\omega, hg^{-1}b, h\beta(t))\lambda(dh)\Delta(g^{-1})\kappa_{\beta(t),t}(dg)\xi(\omega, dt)\mathbb{Q}(d(\omega, b)),$$

by Fubini's theorem and the characteristic property (2.1) of the modular function. Now apply (4.36) to the function $(\omega, s, t) \mapsto \int f(\theta_h\omega, hs, ht)\lambda(dh)$ to write this as

$$\iiint f(\theta_h, ht, hb)\lambda(dh)\xi(\omega, dt)\mathbb{Q}(d(\omega, b)) = \iiint f(\theta_h, t, hb)\lambda(dh)\xi(\theta_h\omega, dt)\mathbb{Q}(d(\omega, b)),$$

and by Fubini's theorem and the definition of C this is just

$$\iint f(\omega, t, s)\xi(dt)C(d(\omega, s)).$$

Hence $\mathbb{Q}$ satisfies (4.2) and since it also satisfies (4.3) by assumption the uniqueness from Theorem 4.1 implies that $\mathbb{Q}$ must be the Palm measure of ξ with respect to O and $\mathbb{P}$.   □

The striking feature of the above Theorem 4.33 is that it is entirely intrinsic, i.e. no other objects than ξ, O and $\mathbb{Q}$ are needed in order to check whether $\mathbb{Q}$ is the cumulative Palm measure of ξ with respect to O. In view of (4.2) it seems surprising at first sight that $\mathbb{P}$ does not play a role at all in this characterization. But the fact that $\mathbb{P}$ may be reconstructed in large parts only by means of $\mathbb{Q}$ (see Lemma 4.9) makes it plausible that an intrinsic characterization of $\mathbb{Q}$ as above is possible.

Chapter 5

The Mass-Transport Principle

The *Mass-Transport Principle (MTP)* is a simple (deterministic) statement about jointly G-invariant measures on a product space $S \times T$ and has been successfully employed in various stationary models in Probability Theory. Early versions have been used by Liggett [43], Adams [1] and van den Berg and Meester [70]. Häggström [24] was the first who successfully applied it in percolation theory. It then became an indispensable tool in this field, see [6, 7] and also [44]. Last and Thorisson [39] and Last [37, 38] derived an MTP for a special class of jointly G-stationary random measures on $G \times G$ by specializing Neveu's classical *exchange formula* in [56] (see our generalization here in Theorem 4.27). Before introducing it in its greatest generality in Section 5.2, we will first give motivations in special cases in Section 5.1. These will clarify necessary ingredients in the fully general (possibly non-transitive and possibly non-unimodular) case. In Section 5.3 we shall then show how this deterministic principle may be applied to produce results about random measures and transports. The final Section 5.4 is devoted to an application of our general MTP to automorphism-stationary random subgraphs (e.g. subgraphs resulting from an automorphism invariant percolation model) of possibly non-unimodular and non-transitive graphs. We shall relate the distributions of various *typical clusters* with the distributions of various 0-*clusters*. Later, in Chapter 7, we shall also give applications of our new form of the MTP to spatial processes on manifolds and to the approximation of Borel sets.

5.1 Motivations

In this section we consider a lcsc group G with Haar measure λ operating measurably on a measurable space S and investigate properties of jointly G-invariant measures on the product space $S^2 = S \times S$ in various special cases before proceeding with the most general situation in Section 5.2. By *Mass-Transport Principle* we mean an equality of the type

$$M(B \times S) = M(S \times B), \tag{5.1}$$

where the intuition of transporting mass comes from the fact that for $C, D \subset S$ the quantity $M(C \times D)$ may be interpreted as mass transported out of the region C into the region D. The reader should convince herself at this point that the defining properties of a measure M on a product space really mean nothing but the physical realizability of a transportation plan where $M(C \times D)$ gives the information how much mass is transported out of $C \subset S$ into $D \subset S$. It will turn out that the mass conservation law (5.1), suitably modified if the operating group G is not unimodular, holds for certain sets $B \subset S$ as a consequence of the joint G-stationary by a simple change in the order of summation resp. integration. As this idea becomes most transparent in the simplest possible setting, namely that of a discrete group operating on itself, we shall focus our attention on this special case first.

5.1.1   A transitive unimodular case

Consider $\mathbb{Z}^2$ operating on itself via translation and a jointly $\mathbb{Z}^2$-invariant measure M on $\mathbb{Z}^2 \times \mathbb{Z}^2$. As we work in a discrete setting the joint $\mathbb{Z}^2$-invariance may be equivalently rephrased by the pointwise property

$$M\{(g+s, g+t)\} = M\{(s,t)\}, \quad g,s,t \in \mathbb{Z}^2.$$

This means that the amount of mass transported from s to t is the same as that transported from $g+s$ to $g+t$ for arbitrary $g \in \mathbb{Z}^2$ (see Figure 5.1).

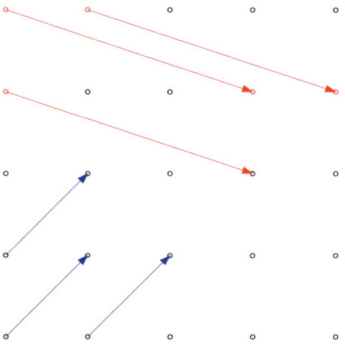

Figure 5.1: Identifying the mass transported from s to t by a pointer starting in s and ending in t the red pointers represent transports with identical masses just as the blue pointers do. The transported mass represented by a red pointer may differ from that represented by a blue pointer.

Fixing a point $b \in \mathbb{Z}^2$ the term $M(\mathbb{Z}^2 \times \{b\})$ then denotes the total mass transported into b while $M(\{b\} \times \mathbb{Z}^2)$ represents the total mass transported out of b. Now we may write $M(\mathbb{Z}^2 \times \{b\})$ as

$$\sum_{z \in \mathbb{Z}^2} M(z, b)$$

and use the joint invariance of M to replace $M(z, b)$ by $M(b, 2b-z)$ (simply add $b-z$ in both components). Since $z \mapsto 2b - z$ is clearly a bijection on $\mathbb{Z}^2$ the sum equals (changing the order of summation) $\sum_{z \in \mathbb{Z}^2} M(b, z)$. We thus proved the conservation law

$$M(\{b\} \times \mathbb{Z}^2) = M(\mathbb{Z}^2 \times \{b\}), \quad b \in \mathbb{Z}^2,$$

(see Figure 5.2 for an illustration of a more general statement) which clearly extends to arbitrary subsets $B \subset \mathbb{Z}^2$ by σ-additivity of M:

$$M(B \times \mathbb{Z}^2) = M(\mathbb{Z}^2 \times B), \quad B \subset \mathbb{Z}^2.$$

Hence for any subset $B \subset \mathbb{Z}^2$ *the total mass transported out of B equals the total mass transported into B*. Note that no extra requirements on the set B are necessary here. This changes completely in a non-transitive setting as we shall see in Subsection 5.1.3. In addition it should be clear from the above arguments and Figure 5.2 that even

$$M(B \times \mathbb{Z}^2) = M(\mathbb{Z}^2 \times C), \quad B, C \subset \mathbb{Z}^2,$$

whenever $|B| = |C|$. Note that in this setting $|B|$ is nothing but the width $\delta(B)$ of B defined as in (2.12).

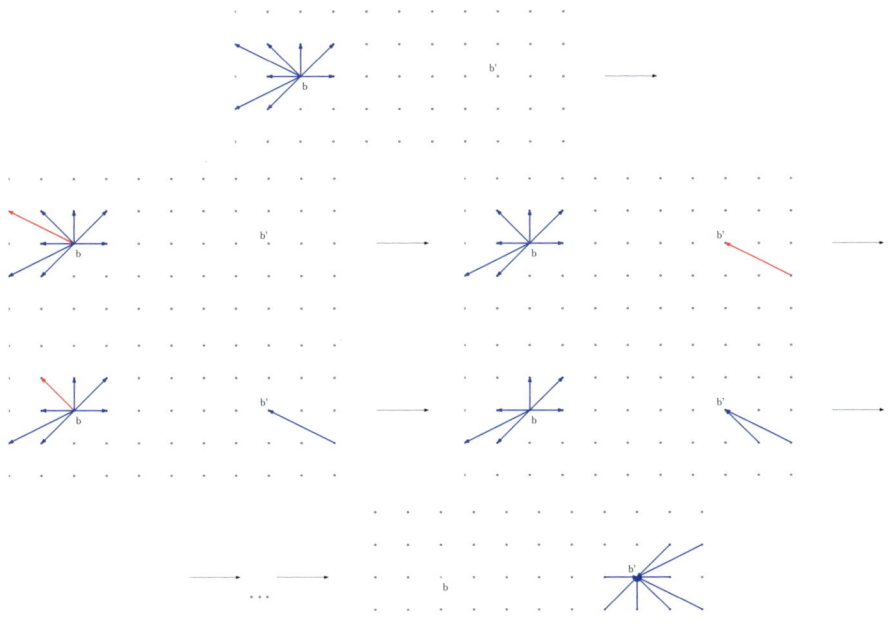

Figure 5.2: The total mass transported out of a point $b \in \mathbb{Z}^2$ equals the total mass transported into a possibly different point $b' \in \mathbb{Z}^2$.

5.1.2 A transitive non-unimodular case

Of particular interest to us are automorphism groups of graphs. Comprehensive treatments of Graph Theory may be found e.g. in [8, 17]. We only recall very few basic notions from this realm here with an emphasis on the topological properties of graph automorphism groups.

A *graph* is a pair $\Gamma := (V, E)$ where V is any set and E any *symmetric* subset of $V \times V$, where symmetry means that $(x, y) \in E \Leftrightarrow (y, x) \in E$. (More precisely these are the *unoriented* or *undirected* graphs). If V is countable, then Γ is also called *countable*. An element $v \in V$ is called *vertex* while $e \in E$ is called an *edge*, where we abbreviate an edge (x, y) by xy. A *neighbor* of a vertex v is another vertex $w \in V$ such that $vw \in E$. The *degree* $\deg(v)$ of a vertex v is the number of its different neighbors. A *path* is a non-empty graph $P = (V_P, E_P)$ of the form

$$V_P = \{x_0, x_1, x_2, \dots\} \quad \text{and} \quad E_P = \{x_0 x_1, x_1 x_2, \dots\},$$

where the x_i are all distinct. Paths may have finite of infinite length. We write a path P of length $k \geq 1$ as $P =: x_0 x_1 \dots x_k$, i.e. in terms of its sequence of vertices. A *ray* is a path which starts in a given vertex x and is infinite in the other direction. Two rays are *equivalent* if they share all but finitely many vertices. If

$$P = (V_P = \{x_0, x_1, \dots, x_k\}, E_p = \{x_0 x_1, x_1 x_2, \dots, x_{k-1} x_k\})$$

is a path then $C := (V_P, E_P \cup \{x_k x_0\})$ is called a *cycle*. A *forest* is a graph that does not contain cycles, while a *tree* is a connected graph without cycles. Two graphs $\Gamma = (V, E)$ and $\Gamma' = (V', E')$ are called *isomorphic* if there is a bijection $\varphi : V \to V'$ satisfying

$$xy \in E \quad \Leftrightarrow \quad \varphi(x)\varphi(y) \in E'.$$

Such a φ is called *isomorphism* between Γ and Γ'. If $\Gamma = \Gamma'$ then φ is called *automorphism* of Γ. The set of these automorphisms clearly forms a group which we denote by $\mathrm{Aut}(\Gamma)$. It is easy to see that a map $\varphi : V \to V$ is a graph automorphism of Γ if and only if it is an isometry with respect to the natural discrete metric $d : V \times V \to [0, \infty)$ which measures the distance between $x, y \in V$ in terms of the number of edges a shortest path connecting x and y has.

The automorphism group $G := \mathrm{Aut}(\Gamma)$ of a countable graph $\Gamma = (V, E)$ is endowed with the topology of pointwise convergence, where V is given the discrete topology. Thus

$$G \ni \varphi_n \to \varphi \in G \quad \Leftrightarrow \quad \forall x \in V : \varphi_n(x) = \varphi(x) \text{ for all but finitely many } n \in \mathbb{N}.$$

With this topology the stabilizers $G_{x,x}, x \in V$, are both open and compact [69, 73] and the family of (also open and compact) sets $G_{x,y}, x, y \in V$, is a (countable) subbase of the topology on G [69, 73], thus G is second countable. It is easy to see that this topology is Hausdorff (and totally disconnected). In addition, as any φ evidently lies in some $G_{x,y}$ for suitable $x, y \in V$ the group G is locally compact. Summarizing, G is lcsc with this topology and thus carries a σ-finite Haar measure λ. The natural operation of G on V given by

$$(\varphi, x) \mapsto \varphi(x)$$

will be denoted by $G \hookrightarrow V$ and is continuous as the preimage of a vertex $y \in V$ under the above map (V carries the discrete topology) equals

$$\{(\varphi, x) : \varphi(x) = y\} = \bigcup_{x \in V} (G_{x,y} \times \{x\})$$

which is clearly open in $G \times V$. In particular $G \hookrightarrow V$ is measurable. Further since $\pi_v^{-1}(\{w\}) = G_{v,w}, v, w \in V$, is compact also $\pi_v^{-1}(K), v \in V$, is compact for any compact (i.e. finite) $K \subset V$. Thus $G \hookrightarrow V$ is topologically proper, as claimed in Subsection 2.2.3. In particular it is proper in the wider sense which is also compatible with Lemma 3.8 since $\emptyset \neq G_{s,s}$ is open and compact and thus $0 < \lambda(G_{s,s}) < \infty, s \in V$.

A countable graph Γ is called *(vertex) transitive* if $\mathrm{Aut}(\Gamma) \hookrightarrow V$ is transitive and *quasi-transitive* if there are at most finitely many orbits. Further, Γ is called *unimodular* if $\mathrm{Aut}(\Gamma)$ is unimodular.

Turning to the promised example, we denote for $n \geq 2$ the up to isomorphy unique countable tree in which each vertex has degree n by T_n, see Figure 5.3 for a picture of T_3. It is clear from Lemma 3.17 in combination with Lemma 3.13 that T_n is unimodular. An *end* of a tree is an equivalence class of rays. The following construction (essentially) stems from Trofimov [69]. Let $n \geq 3$. Given an end ξ in the n-regular tree $T_n = (V, E)$ then for each $x \in V$ there is clearly a unique ray x_ξ of the form $x x_1 x_2 \ldots$ (i.e. starting in x) lying in the equivalence class ξ (also drafted in Figure 5.3). In this situation $x_2 =: \xi(x)$ is called ξ-*grandparent* of x. Note

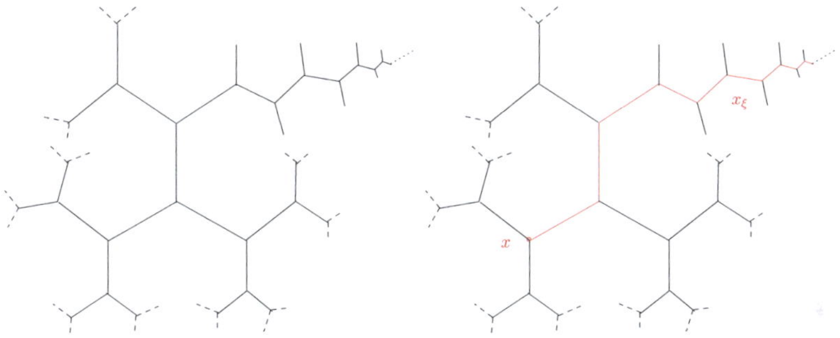

Figure 5.3: Scheme of T_3 and a ray $x_\xi \in \xi$ starting in x.

that in the case of T_3 each vertex is the ξ-grandfather of 4 different vertices which generalizes to T_n where each vertex is the ξ-grandfather of $(n-1)^2$ different vertices (its *grandchildren*). Adding new edges between $\xi(x)$ and each of its grandchildren gives the graph indicated in Figure 5.4 on the left, and repeating this step for each vertex finally gives the ξ-*grandparent graph on* T_n (i.e. after adding an edge between each vertex x and its grandparent $\xi(x)$). We call it $\xi(T_n)$.

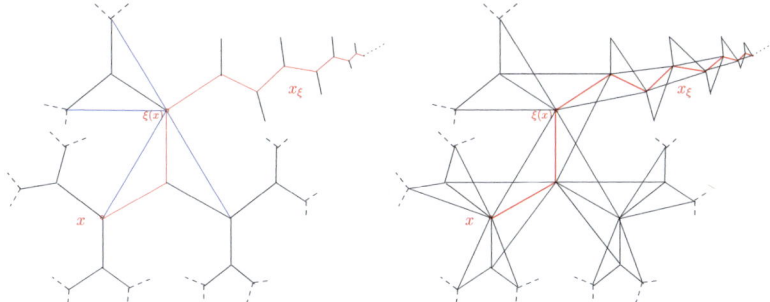

Figure 5.4: Left: $\xi(x)$ connected by an edge with each of its grandchildren, right: $\xi(T_3)$.

It is easy to see that $\xi(T_n)$ is transitive and that each $\varphi \in \mathrm{Aut}(\xi(T_n))$ fixes ξ in the sense that for each ray $\rho \in \xi$ again $\varphi(\rho) \in \xi$ (where $\varphi(\rho)$ is defined in the obvious way). Thus the map

$$m(s,t) := \mathbf{1}\{t = \xi(s)\}, \quad s,t \in V,$$

is jointly $\mathrm{Aut}(\xi(T_n))$-invariant which is then also true for the measure

$$M(\cdot) := \sum_{s,t \in V} \mathbf{1}\{(s,t) \in \cdot\} \mathbf{1}\{t = \xi(s)\}$$

on $V \times V$ (M governs the transport in which each vertex sends mass 1 to its grandparent). Here a conservation law of the type (5.1) fails for *any* subset $B \subset V$ since for any vertex $o \in V$

$$M(\{o\} \times V) = \sum_{t \in V} \mathbf{1}\{t = \xi(o)\} = 1 < (n-1)^2 = \sum_{s \in V} \mathbf{1}\{o = \xi(s)\} = M(V \times \{o\}).$$

This reasoning is taken from [44, p. 206]. The crucial difference to the transitive unimodular situation in Subsection 5.1.1 is that $\xi(T_n)$ is *not* unimodular (i.e. the operating group $\mathrm{Aut}(\xi(T_n))$ is not). This readily follows from Lemma 3.17 in combination with Lemma 3.13, since for any $v \in V$ evidently $|G_{v,v}\xi(v)| = 1$, while $|G_{\xi(v),\xi(v)}v| = 4$. In this context, we also note that we shall compute the non-trivial functions Δ^* with respect to a fixed representative o, and $\tilde{\Delta}$ in Subsection 5.4.2. As this example shows, the topology of the graph, i.e. that of the operating group of automorphisms has an effect. As we show in Section 5.2, this effect *only* stems from the non-trivial modular function of the operating group and we shall provide a suitable counterbalancing density that rescues a conservation law even in a possibly non-unimodular and even a possibly non-transitive setting. The differences arising from non-transitivity are motivated in the next subsection.

5.1.3   A non-transitive case

Exemplarily we consider here the evidently non-transitive operation $SO(2) \hookrightarrow \mathbb{R}^2$ where the orbit of a point $b \in \mathbb{R}^2$ is the circle around the origin containing b. Since $SO(2)$ is compact we may normalize Haar measure on $SO(2)$ to a probability measure which we call λ. We choose an arbitrary but fixed measurable system O of representatives of the orbits, e.g. $O = \{(x,0) : x \geq 0\}$. Then for $b \in O$ the orbital pushforward $\mu_b = \lambda \circ \pi_b^{-1}$ is the uniform distribution on the circle around the origin through b. In this setting the above conservation law fails in general for arbitrary subsets $B \in \mathcal{B}(\mathbb{R}^2)$. Consider for instance the jointly $SO(2)$-invariant measure $M = \mu_b \otimes \mu_c$ for $b,c \in \mathbb{R}^2$ not lying in the same orbit. Here clearly (since $\lambda(SO(2)) = 1$)

$$M(B \times \mathbb{R}^2) = M(\mathbb{R}^2 \times B) \quad \Leftrightarrow \quad \mu_b(B) = \mu_c(B),$$

and hence if

$$M(B \times \mathbb{R}^2) = M(\mathbb{R}^2 \times B)$$

is to be satisfied for *all* jointly G-invariant measures M then sending (b, c) through all different pairs of orbit representatives, we get as a necessary condition on B that

$$\mu_b(B) = \mu_c(B), \quad b, c \in O.$$

Thus, B must be $SO(2)$-symmetric (recall the examples in Figure 2.1). It turns out that this necessary condition on B is already sufficient for a mass-conservation law of the above form (see Corollary 5.4) such that

$$M(B \times \mathbb{R}^2) = M(\mathbb{R}^2 \times B)$$

holds for all jointly $SO(2)$-invariant measures M on $\mathbb{R}^2 \times \mathbb{R}^2$ if and only if B is $SO(2)$-symmetric.

5.2 The Mass-Transport Principle

In this section we state and prove two forms of the mass-transport principle (MTP) for possibly non-unimodular operating groups and possibly non-transitive operations. The first form is given in Subsection 5.2.1 and represents a mass-conservation law on systems of orbit representatives. This first version is then needed in Subsection 5.2.2 in the proof of the second, 'integrated' version in Theorem 5.2, which we simply call *the* Mass-Transport Principle.

5.2.1 MTP on systems of orbit representatives

We consider the proper action of an lcsc group G on a Borel space S and a Borel space T. Similarly as before we ask for necessary and sufficient conditions on subsets B and C such that a mass-conservation law (suitably modified in the non-unimodular case) is fulfilled for all jointly G-invariant σ-finite (or s-finite) measures M on $S \times T$. Similarly as in the discrete transitive example of Subsection 5.1.1 we first need to establish an MTP on a system of representatives before integrating it to a version on G-symmetric sets.

Lemma 5.1 (MTP on representatives). *Let G operate properly on the Borel spaces S and T, let μ and ν denote G-invariant σ-finite measures on S and T respectively, and let γ and δ denote G-invariant s-finite kernels from S to T and T to S respectively. If*

$$\iint \mathbf{1}\{(s, t) \in \cdot\}\gamma(s, dt)\mu(ds) = \iint \mathbf{1}\{(s, t) \in \cdot\}\delta(t, ds)\nu(dt) \qquad (5.2)$$

holds, then we have for any jointly G-invariant function $m : S \times T \to [0, \infty]$ that

$$\iint m(s, b)\delta(b, ds)\nu^*(db) = \iint m(b, t)\Delta^*(t)\gamma(b, dt)\mu^*(db). \qquad (5.3)$$

Proof. Choose $w : S \to (0, \infty)$ as in (2.6). Then the left-hand side may be written as

$$\iint m(s, b)\delta(b, ds)\nu^*(db) = \iiint m(s, \beta(t))\delta(\beta(t), ds)w(t)\mu_b(dt)\nu^*(db)$$

which equals by (2.7)

$$\iint m(s, \beta(t))w(t)\delta(\beta(t), ds)\nu(dt).$$

Using the inversion kernel κ of the operation $G \hookrightarrow T$ we may write this as

$$\iiint m(s, h^{-1}t)w(t)\delta(h^{-1}t, ds)\kappa_{\beta(t),t}(dh)\nu(dt)$$
$$= \iiint m(h^{-1}s, h^{-1}t)w(t)\delta(t, ds)\kappa_{\beta(t),t}(dh)\nu(dt).$$

By the joint invariance of m and (5.2) this reduces to

$$\iint m(s, t)w(t)\delta(t, ds)\nu(dt) = \iint m(s, t)w(t)\gamma(s, dt)\mu(ds).$$

We now reverse the above steps. Applying (2.7) yields

$$\iiint m(s, t)w(t)\gamma(s, dt)\mu_b(ds)\mu^*(db) = \iiint m(gb, t)w(t)\gamma(gb, dt)\lambda(dg)\mu^*(db)$$

and using invariance of γ and joint invariance of m we arrive again by (2.6) at

$$\iiint m(b, t)w(gt)\gamma(b, dt)\lambda(dg)\mu^*(db) = \iint m(b, t)\mu_t(w)\gamma(b, dt)\mu^*(db).$$

This yields the assertion since $\mu_t(w) = \Delta^*(t)$ by Lemma 3.12. $\qquad\square$

5.2.2   Integrated version

Lemma 5.1 will be needed to derive the following theorem. Recall the definition of $\tilde{\Delta}$ in (3.12) and its properties from Lemma 3.14.

Theorem 5.2 (Mass-Transport Principle). *Let G operate properly on the Borel spaces S and T and consider non-negative functions k^S on S and k^T on T. Then*

$$\int k^S(s)\tilde{\Delta}(s, t)M(d(s, t)) = \int k^T(t)M(d(s, t)) \tag{5.4}$$

for all σ-finite jointly G-invariant measures M on $S \times T$ if and only if

$$\mu_b k^S = \mu_c k^T, \quad b \in O_S, c \in O_T. \tag{5.5}$$

Here the word σ-finite may be replaced by s-finite.

Proof. First assume that k^S, k^T fulfill (5.5) and take a σ-finite jointly G-invariant measure M on $S \times T$. Then by Theorem 2.16 there exist both an invariant disintegration from S to T

$$M(d(s, t)) = \gamma(s, dt)\mu(ds)$$

and an invariant disintegration from T to S

$$M(d(s,t)) = \delta(t, ds)\nu(dt).$$

Then (5.4) may be written as

$$\iint \tilde{\Delta}(s,t)k^S(s)\gamma(s, dt)\mu(ds) = \iint k^T(t)\delta(t, ds)\nu(dt)$$

and employing (2.7) an equivalent statement is

$$\iiint \tilde{\Delta}(gb,t)k^S(gb)\gamma(gb, dt)\lambda(dg)\mu^*(db) = \iiint k^T(gc)\delta(gc, ds)\lambda(dg)\nu^*(dc).$$

Using invariance of the respective kernels this may be stated equivalently as

$$\iiint \tilde{\Delta}(gb,gt)k^S(gb)\gamma(b, dt)\lambda(dg)\mu^*(db) = \iiint k^T(gc)\delta(c, ds)\lambda(dg)\nu^*(dc).$$

Since $\tilde{\Delta}$ is jointly G-invariant we may cancel by (5.5) the identical constants $\mu_b k^S = \mu_c k^T$ on both sides and arrive, since $\tilde{\Delta}(b,t) = \Delta^*(t)$, at (5.3) in its form for $m \equiv 1$, which is true by Lemma 5.1.

Conversely, assume (5.4) for all jointly G-invariant M. Consider for fixed $b \in O_S, c \in O_T$ the jointly G-invariant σ-finite measure

$$M := \int \mathbf{1}\{(hb, hc) \in \cdot\}\lambda(dh),$$

which admits the following two disintegrations in opposite directions

$$M = \iint \mathbf{1}\{(s, gc) \in \cdot\}\kappa_{b,s}(dg)\mu_b(ds) = \iint \mathbf{1}\{(gb, t) \in \cdot\}\kappa_{c,t}(dg)\mu_c(dt)$$

(this may be checked by direct calculation, each disintegration reduces to the definition of M by using right $G_{b,b}$-invariance of λ, i.e. (2.5)). Using the left disintegration on the left side of (5.4) and the right disintegration on the right side of (5.4) yields

$$\iint k^S(s)\tilde{\Delta}(s, gc)\kappa_{b,s}(dg)\mu_b(ds) = \iint k^T(t)\kappa_{c,t}(dg)\mu_c(dt).$$

While the right side clearly equals $\mu_c k^T$, the left side requires a bit of manipulation. First

$$\iint k^S(s)\tilde{\Delta}(s, gc)\kappa_{b,s}(dg)\mu_b(ds) = \iint k^S(hb)\tilde{\Delta}(hb, gc)\kappa_{b,hb}(dg)\lambda(dh)$$

$$= \iint k^S(hb)\tilde{\Delta}(hb, hgc)\kappa_{b,b}(dg)\lambda(dh)$$

and using joint G-invariance of $\tilde{\Delta}$ this reduces to

$$\mu_b k^S \int \tilde{\Delta}(b, gc)\kappa_{b,b}(dg).$$

Again by joint G-invariance of $\tilde{\Delta}$ this equals

$$\mu_b k^S \int \tilde{\Delta}(g^{-1}b, c)\kappa_{b,b}(dg),$$

which clearly reduces to $\mu_b k^S \tilde{\Delta}(b, c)$. Here $\tilde{\Delta}(b, c) = 1$ by (3.13) which yields the assertion. $\qquad\square$

The special case $S = T$ and $k^S = k^T$ have been treated in [21] by specializing a version of the Transport Theorem 4.27 for $S = T$. Instead of using the Transport Theorem 4.27 in order to derive Lemma 5.1 we chose to give direct proofs here because of the resulting increase of transparency about how the joint invariance is used. Using G-symmetric sets (i.e. sets B with $0 < \mu_b(B) = \mu_c(B) < \infty, b, c \in O$) in both spaces S and T the Mass-Transport Theorem 5.2 may be stated as follows:

Theorem 5.3 (Mass-Transport Principle on sets). *Let G operate properly on the Borel spaces S and T and consider sets $B \in \mathcal{S}$ and $C \in \mathcal{T}$. Then*

$$\int \mathbf{1}_B(s)\tilde{\Delta}(s,t)M(d(s,t)) = M(S \times C) \qquad (5.6)$$

for all σ-finite jointly G-invariant measures M on $S \times T$ if and only if B and C are both G-symmetric and have the same width, i.e.

$$\delta(B) = \delta(C). \qquad (5.7)$$

Here the word σ-finite may be replaced by s-finite.

We will mainly use this 'set formulation' of the MTP and again mainly in the further specialization $S = T$, $B = C$:

Corollary 5.4 (Mass-Transport Principle on one set). *Let G operate properly on the Borel space S and consider a set $B \in \mathcal{S}$. Then*

$$\int \mathbf{1}_B(s)\tilde{\Delta}(s,t)M(d(s,t)) = M(S \times B) \qquad (5.8)$$

for all σ-finite jointly G-invariant measures M on $S \times T$ if and only if B is G-symmetric. Here the word σ-finite may be replaced by s-finite.

5.3   MTP's for random measures

We will mostly use the above Mass-Transport Principle in order to relate distributions of certain suitably invariant random elements. In this section we show the relation between G-stationary random elements and the deterministic MTP.

5.3.1   MTP and stationary random measures

Consider a random measure ζ on $S \times T$. Similarly as in the deterministic case, where we interpreted $M(C \times D)$ as mass transported out of C into D, we may interpret $\zeta(C \times D)$ as random mass transported out of C into D. Note that the intensity measure of $C \times D$, $\mathbb{E}\zeta(C \times D)$, may be interpreted as the expected transported mass from C to D. Since we introduced random measures as σ-finite kernels, $\mathbb{E}\zeta$ is s-finite by Lemma 2.20. If in addition ζ is G-stationary, then $\mathbb{E}\zeta$ is jointly G-invariant:

$$\mathbb{E}\int f(gs, gt)\zeta(d(s,t)) = \mathbb{E}\int f(s,t)\zeta(\theta_g, d(s,t)) = \mathbb{E}\int f(s,t)\zeta(d(s,t)), \quad f \in (\mathcal{S} \otimes \mathcal{T})_+.$$

Here we used G-invariance of $\mathbb{P}$, which we may and will assume without loss of generality, see Subsection 2.4.2. An application of one direction of the set version of the deterministic MTP (Theorem 5.3) yields

Theorem 5.5 (MTP for random transports). *Let ζ denote a G-stationary random measure on $S \times T$ and let $B \in \mathcal{S}$ and $C \in \mathcal{T}$ denote G-symmetric sets with the same width. Then*

$$\mathbb{E} \int \mathbf{1}_B(s)\tilde{\Delta}(s,t)\zeta(d(s,t)) = \mathbb{E}\zeta(S \times C). \tag{5.9}$$

and for any jointly G-invariant $m : \Omega \times S \times T \to [0,\infty)$

$$\mathbb{E} \int \mathbf{1}_B(s)\tilde{\Delta}(s,t)m(\theta_e,s,t)\zeta(d(s,t)) = \mathbb{E} \int \mathbf{1}_C(t)m(\theta_e,s,t)\zeta(d(s,t)). \tag{5.10}$$

Proof. The first equation is clear after what has been said above the theorem. To prove the second we may simply apply (5.9) to the G-stationary random measure

$$\zeta'(\omega, d(s,t)) := m(\omega,s,t)\zeta(\omega, d(s,t)),$$

and refer to Lemma 2.21 (i). $\qquad\square$

Note that if ξ is a G-stationary random measure on S and γ a G-invariant kernel from $\Omega \times S$ to T then

$$\zeta(\omega, \cdot) := (\xi \otimes \gamma)(\omega)$$

is a suitable choice for ζ in the above theorem. The same remark applies to the product of a G-stationary random measure η on T and a G-invariant kernel δ from $\Omega \times T$ to S. Clearly if $\xi \otimes \gamma = \eta \otimes \delta$ $\mathbb{P}$-a.e. then

$$\mathbb{E} \iint \mathbf{1}_B(s)\tilde{\Delta}(s,t)m(\theta_e,s,t)\gamma(s,dt)\xi(ds) = \mathbb{E} \iint \mathbf{1}_C(t)m(\theta_e,s,t)\delta(t,ds)\eta(dt). \tag{5.11}$$

For random transports ζ in this form and the special choice of $S = T, B = C$ Theorem 5.5 has been shown in [21, Theorem 5] by using a special form of the Transport Theorem 4.27. The next section will clarify the link between MTP and Transport Formula.

5.3.2 Palm Mass-Transport Principle

There is a close link between the MTP in the form of (5.11) and the Transport Theorem in the special form of equation (4.31). Equation (5.11) may be rewritten by means of the cumulative Palm measures of ξ and η by applying (4.2) in its respective form on either of the two sides of (5.11). This yields

$$\iiint \mathbf{1}_B(gb)\tilde{\Delta}(gb,t)m(\theta_g\omega,gb,t)\gamma(\theta_g\omega,gb,dt)\lambda(dg)\mathbb{Q}^\xi(d(\omega,b))$$
$$= \iiint \mathbf{1}_C(gb)m(\theta_g\omega,s,gb)\delta(\theta_g\omega,gb,ds)\lambda(dg)\mathbb{Q}^\eta(d(\omega,b)),$$

and using invariance of γ and δ, joint invariance of m and $\tilde{\Delta}$ and Fubini we arrive at

$$\iiint \mathbf{1}_B(gb)\lambda(dg)\tilde{\Delta}(b,t)m(\omega,b,t)\gamma(\omega,b,dt)\mathbb{Q}^\xi(d(\omega,b))$$
$$= \iiint \mathbf{1}_C(gb)\lambda(dg)m(\omega,s,b)\delta(\omega,b,ds)\mathbb{Q}^\eta(d(\omega,b)).$$

Canceling the identical constants $\delta(B) = \delta(C)$ on both sides yields

Theorem 5.6 (Palm MTP). *Let ξ and η denote G-stationary random measures on S and T respectively and let γ and δ be G-invariant kernels from $\Omega \times S$ to T resp. $\Omega \times T$ to S such that*

$$\iint \mathbf{1}\{(s,t) \in \cdot\}\gamma(s, dt)\xi(ds) = \iint \mathbf{1}\{(s,t) \in \cdot\}\delta(t, ds)\eta(dt) \quad \mathbb{P}\text{-}a.e. \qquad (5.12)$$

Then we have for any jointly G-invariant $m : \Omega \times S \times T \to [0, \infty)$ that

$$\iint \Delta^*(t)m(\omega, b, t)\gamma(\omega, b, dt)\mathbb{Q}^\xi(d(\omega, b)) = \iint m(\omega, s, b)\delta(\omega, b, ds)\mathbb{Q}^\eta(d(\omega, b)). \qquad (5.13)$$

If only the weaker condition

$$\mathbb{E} \iint \mathbf{1}\{(s,t) \in \cdot\}\gamma(s, dt)\xi(ds) = \mathbb{E} \iint \mathbf{1}\{(s,t) \in \cdot\}\delta(t, ds)\eta(dt) \qquad (5.14)$$

holds, then

$$\iint \Delta^*(t)\gamma(\omega, b, dt)\mathbb{Q}^\xi(d(\omega, b)) = \iint \delta(\omega, b, ds)\mathbb{Q}^\eta(d(\omega, b)). \qquad (5.15)$$

Remark 5.7. Note that the existence of G-symmetric subsets B and C is open in this generality. Theorem 5.6 is still valid as we may repeat the same calculation using the (existent) functions w^S and w^T as in (2.6) this time starting with the obvious variant of (5.11) using symmetric functions instead of sets. Another possibility is to specialize (4.31). Replacing $f := m$ and using its joint G-invariance clearly immediately yields Theorem 5.6. Thus there is an intimate link between the Transport Theorem 4.27 and the Mass Transport principle in the form of Theorem 5.6. Instead of going this shorter way we chose to give the above proof using the MTP since the intuition of transporting mass behind the MTP is useful in applications.

5.4   Application: Stationary subgraphs

Most results derived from the Mass-Transport Principle are qualitative. As an illustration of how the MTP works in its most general form we decided to work here in a setting which is not necessarily transitive nor necessarily unimodular. For the time being, we want our system of orbit representatives O to be compact since then there are good chances that $0 < (\mathbb{E}\xi)^*(O) < \infty$ is fulfilled which according to Subsection 4.3.1 will allow probabilistic interpretations without further modifications (such as introducing invariant sets with 'finite height'). Thus a suitable setting certainly is a quasi-transitive not necessarily unimodular graph Γ and a suitable random object is an $\mathrm{Aut}(\Gamma)$-stationary random subgraph, e.g. the result of an $\mathrm{Aut}(\Gamma)$-stationary bond percolation, see e.g. [9, 22]. Such models are discussed in Subsection 5.4.1. The result obtained there will be made more explicit in a special transitive case, namely on $\xi(T_n)$ in Subsection 5.4.2.

5.4.1 Stationary subgraphs in quasi-transitive graphs

Let $\Gamma = (V, E)$ denote a quasi-transitive possibly non-unimodular graph and $G :=$ $\mathrm{Aut}(\Gamma)$ the lcsc group of automorphisms (see Subsection 5.1.2) on Γ with a fixed left Haar measure λ. We consider the continuous and topologically proper (again see Subsection 5.1.2) natural operation $G \hookrightarrow V$ given by $(\varphi, v) \mapsto \varphi(v)$ and O shall denote a fixed (finite) complete system of orbit representatives. We denote by Ξ the set of all nonempty subgraphs of Γ and endow Ξ with the σ-algebra generated by the evaluation maps

$$\rho_v : \Xi \to \{0, 1\}, \quad \rho_v(H) := \mathbf{1}\{v \in V(H)\}, \quad v \in V,$$

and

$$\rho_e : \Xi \to \{0, 1\}, \quad \rho_e(H) := \mathbf{1}\{e \in E(H)\}, \quad e \in E.$$

The operation $G \hookrightarrow V$ induces naturally an operation $G \hookrightarrow \Xi$ such that we may write for $H = (V', E')$ the shifted graph $(\varphi(V'), \varphi(E'))$, where $\varphi(V') := \{\varphi(v) : v \in V'\}$ and $\varphi(E') := \{\varphi(v_1)\varphi(v_2) : v_1 v_2 \in E'\}$, simply as φH. Then a *random subgraph* ϑ of Γ is a random element in Ξ defined on an underlying probability space $(\Omega, \mathcal{A}, \mathbb{P})$. As explained in Subsection 2.4.2 we may model stationarity without loss of generality by equipping Ω with a flow θ indexed by $\mathrm{Aut}(\Gamma)$, by assuming that $\mathbb{P}$ is invariant with respect to this flow, and by adapting our G-stationary random elements to this flow. Thus, a random subgraph ϑ of Γ is $G = \mathrm{Aut}(\Gamma)$-stationary, if

$$\vartheta(\theta_\varphi \omega) = \varphi \vartheta(\omega), \quad \omega \in \Omega, \varphi \in G.$$

A *cluster* in a subgraph of Γ is a connected component in this graph and we write $C(\omega, v)$ for the cluster in $\vartheta(\omega)$ that contains v. Let $P(v, w)$ denote the countable set of all paths in Γ connecting the vertices v and w. Since for fixed $v \in V$ the maps

$$\rho_w \circ C(v) = \mathbf{1}\{w \in C(v)\} = \mathbf{1}\{v, w \in V(\vartheta)\} \sup_{P \in P(v,w)} \prod_{e \in P} \mathbf{1}\{e \in E(\vartheta)\}, \quad v \in V,$$

are measurable and since similarly the maps $\rho_e \circ C(v), e \in E$, are measurable, we conclude that $C(v)$ is a random element in Ξ for each $v \in V$. Thus we may investigate distributional properties of such clusters. Further $|H|$ denotes the number of vertices of a subgraph H. Since $H \mapsto |H| = \sum_{w \in V} \mathbf{1}\{w \in H\}$ is evidently measurable we have that $|C(v)|$ is for each $v \in V$ an $\mathbb{N} \cup \{\infty\}$-valued random variable. We make the assumption that

$$\mathbb{E}|C(v)| < \infty, \quad v \in V. \tag{5.16}$$

Remark 5.8. Equation (5.16) is e.g. satisfied if ϑ is the random subgraph resulting from independent Bernoulli(p)-bond percolation on Γ whenever $p < 1/(r-1)$ where $r = \max_{v \in O} \deg(v)$ (see [71, Theorem 6.2]).

Let $\mathcal{C}_f$ denote the space of all finite subgraphs of Γ, endowed with the σ-field inherited from Ξ. We assume the existence of a measurable map

$$\pi : \Omega \times \mathcal{C}_f \to V, \quad (\omega, C) \mapsto \pi(\omega, C)$$

such that

$$\pi(\omega, C) \in C, \quad \omega \in \Omega, C \in \mathcal{C}_f, \tag{5.17}$$

and

$$\pi(\theta_\varphi \omega, \varphi C) = \varphi(\pi(\omega, C)), \quad \varphi \in G, \omega \in \Omega, C \in \mathcal{C}_f. \tag{5.18}$$

Since our focus lies on illustrating the use of Theorem 5.6 we shall not be concerned with the construction of this map in this full generality. Instead we will construct such a map for the special case of $\Gamma = \xi(T_n)$ in Subsection 5.4.2, as it is readily available in this special case.

Open problem 5.9. *Derive the existence of a map $\pi : \Omega \times \mathcal{C}_f \to V$ satisfying both (5.17) and (5.18).*

We call such a function *center function*. We shall fix such a function in the following and interpret for $v \in V$ the random vertex $\omega \mapsto \pi(\omega, C(\omega, v))$ as the *center* of $C(v)$ in configuration ω. We abbreviate

$$\pi(\omega, v) := \pi(\omega, C(\omega, v)), \quad \omega \in \Omega, v \in V.$$

By η we denote the deterministic counting measure on V, which is clearly G-invariant since automorphisms are bijections on V. Further we put $\mathcal{C}(\vartheta) := \{C : C \text{ is a cluster of } \vartheta\}$ and define

$$\xi := \sum_{C \in \mathcal{C}(\vartheta)} \delta_{\pi(C)}. \tag{5.19}$$

It is trivial to show that ξ is a G-stationary and by (5.17) ξ is a simple point process on V. Evidently ξ represents a natural mean to count the clusters of ϑ. We clearly have

$$\mu_b = \lambda(G_{b,b}) \sum_{v \in Gb} \delta_v, \quad b \in O,$$

and thus a suitable function w in (4.4) is $w(v) = \mathbf{1}\{v \in O\}/\lambda(G_{\beta(v),\beta(v)})$. It is easy to derive from (4.4) (using Fubini, invariance of $\mathbb{P}$ and (2.7) for η) or simply from (4.10) that

$$\mathbb{Q}^\eta = \mathbb{P} \otimes \eta^*. \tag{5.20}$$

Further we note that (4.9) holds for ξ as well as for η which implies

$$\eta^* = \sum_{b \in O} \frac{1}{\lambda(G_{b,b})} \delta_b. \tag{5.21}$$

The Mass-Transport Principle in the form of Theorem 5.6 yields, suitably applied, the following result for G-stationary random subgraphs ϑ that contain a.s. all vertices, i.e.

$$\mathbb{P}(v \in \vartheta) = 1, \quad v \in V. \tag{5.22}$$

This is e.g. the case for bond percolation models (see [9, 22]).

Theorem 5.10 (typical clusters). *Let ϑ denote an $\mathrm{Aut}(\Gamma)$-stationary random subgraph of a countable quasi-transitive graph Γ satisfying (5.22) such that each of its clusters satisfies (5.16) and let $C(\omega, v)$ denote the cluster of $v \in V$ in $\vartheta(\omega)$, $\omega \in \Omega$. Let further $\pi : \Omega \times \mathcal{C}_f \to V$ denote a center function, ξ be defined as in (5.19), η denote counting measure on V and O denote a complete (finite) system of orbit representatives. Then the equations*

$$\int \mathbf{1}\{C(\omega, b) \in \cdot\}\mathbb{Q}^\xi(d(\omega, b))$$
$$= \sum_{b \in O} \frac{1}{\lambda(G_{b,b})}\mathbb{E}\left[\frac{1}{|C(\pi(b))|}\Delta^*(\pi(b))\int \mathbf{1}\{\varphi^{-1}C(\pi(b)) \in \cdot\}\kappa_{\beta(\pi(b)),\pi(b)}(d\varphi)\right], \tag{5.23}$$

$$\int \mathbf{1}\{C(\omega, b) \in \cdot\}|C(\omega, b)|\mathbb{Q}^\xi(d(\omega, b))$$
$$= \sum_{b \in O} \frac{1}{\lambda(G_{b,b})}\mathbb{E}\left[\Delta^*(\pi(b))\int \mathbf{1}\{\varphi^{-1}C(\pi(b)) \in \cdot\}\kappa_{\beta(\pi(b)),\pi(b)}(d\varphi)\right], \tag{5.24}$$

$$\int \mathbf{1}\{C(\omega, b) \in \cdot\} \sum_{v \in C(\omega,b)} \Delta^*(v)\mathbb{Q}^\xi(d(\omega, b))$$
$$= \sum_{b \in O} \frac{1}{\lambda(G_{b,b})}\mathbb{E}\left[\int \mathbf{1}\{\varphi^{-1}C(\pi(b)) \in \cdot\}\kappa_{\beta(\pi(b)),\pi(b)}(d\varphi)\right], \tag{5.25}$$

$$\int \mathbf{1}\{C(\omega, b) \in \cdot\} \sum_{v \in C(\omega,b)} \lambda(G_{v,v})\mathbb{Q}^\xi(d(\omega, b))$$
$$= \sum_{b \in O} \mathbb{E}\left[\int \mathbf{1}\{\varphi^{-1}C(\pi(b)) \in \cdot\}\kappa_{\beta(\pi(b)),\pi(b)}(d\varphi)\right], \tag{5.26}$$

hold, where $\Delta^*(v) = \frac{\lambda(G_{v,v})}{\lambda(G_{\beta(v),\beta(v)})} = \frac{|G_{v,v}\beta(v)|}{|G_{\beta(v),\beta(v)}v|}, v \in V$, and where all sides of the equations are finite.

Proof. We define the G-invariant kernels γ and δ from $\Omega \times V$ to V respectively via

$$\gamma(\omega, s, \cdot) := \sum_{v \in C(\omega,s)} \delta_v(\cdot), \quad s \in V, \omega \in \Omega,$$

and

$$\delta(\omega, t, \cdot) := \delta_{\pi(\omega,t)}(\cdot), \quad t \in V, \omega \in \Omega.$$

Then clearly $\iint \mathbf{1}\{t \in \cdot\}\gamma(\omega, s, dt)\xi(\omega, ds) = \eta(\cdot)$, which also means that for $\omega \in \Omega$

$$\iint \mathbf{1}\{(s,t) \in \cdot\}\gamma(\omega, s, dt)\xi(\omega, ds) = \iint \mathbf{1}\{(\pi(\omega, t), t) \in \cdot\}\gamma(\omega, s, dt)\xi(\omega, ds)$$
$$= \int \mathbf{1}\{(\pi(\omega, t), t) \in \cdot\}\eta(dt)$$
$$= \iint \mathbf{1}\{(s,t) \in \cdot\}\delta(\omega, t, ds)\eta(dt).$$

Thus ξ, η, γ and δ satisfy (5.12) and the MTP in the form of Theorem 5.6 yields

$$\iint \tilde{\Delta}(b, t)m(\omega, b, t)\gamma(\omega, b, dt)\mathbb{Q}^\xi(d(\omega, b)) = \iint m(\omega, s, b)\delta(\omega, b, ds)\mathbb{Q}^\eta(d(\omega, b)) \tag{5.27}$$

for any jointly G-invariant measurable $m : \Omega \times V \times V \to [0, \infty]$. We note that for any measurable $D \subset \Xi$ the map

$$m_D(\omega, s, t) := \tilde{\Delta}(t, s) \int \mathbf{1}\{\varphi^{-1}C(\omega, s) \in D\}\kappa_{\beta(s),s}(d\varphi)$$

is jointly G-invariant. Applying (5.27) to such an m yields (omitting D)

$$\int \sum_{v \in C(\omega,b)} \int \mathbf{1}\{\varphi^{-1}C(\omega, b) \in \cdot\}\kappa_{b,b}(d\varphi)\mathbb{Q}^\xi(d(\omega, b))$$
$$= \iiint \mathbf{1}\{\varphi^{-1}C(\omega, s) \in \cdot\}\tilde{\Delta}(b, s)\kappa_{\beta(s),s}(d\varphi)\delta_{\pi(\omega,b)}(ds)\mathbb{Q}^\eta(d(\omega, b))$$

and thus using covariance of C and (5.20) the equation

$$\iint |C(\omega, b)|\mathbf{1}\{C(\theta_\varphi^{-1}\omega, b) \in \cdot\}\kappa_{b,b}(d\varphi)\mathbb{Q}^\xi(d(\omega, b))$$
$$= \iiint \mathbf{1}\{\varphi^{-1}C(\omega, s) \in \cdot\}\tilde{\Delta}(b, s)\kappa_{\beta(s),s}(d\varphi)\delta_{\pi(\omega,b)}(ds)\mathbb{Q}^\eta(d(\omega, b)).$$

Here the left-hand side reduces to

$$\int |C(\omega, b)|\mathbf{1}\{C(\omega, b) \in \cdot\}\mathbb{Q}^\xi(d(\omega, b))$$

after replacing $|C(\omega, b)|$ by $|C(\theta_\varphi^{-1}\omega, b)|$ $(= |\varphi^{-1}C(\omega, b)|$ since $\varphi \in G_{b,b})$ and using (4.3). The right-hand side may be written by (5.20) as

$$\mathbb{E}\iint \mathbf{1}\{\varphi^{-1}C(\omega, \pi(\omega, b)) \in \cdot\}\tilde{\Delta}(b, \pi(\omega, b))\kappa_{\beta(\pi(\omega,b)),\pi(\omega,b)}(d\varphi)\eta^*(db).$$

Using Fubini and (5.21) yields

$$\sum_{b \in O} \frac{1}{\lambda(G_{b,b})}\mathbb{E}\left[\tilde{\Delta}(b, \pi(\omega, b))\int \mathbf{1}\{\varphi^{-1}C(\omega, \pi(\omega, b)) \in \cdot\}\kappa_{\beta(\pi(\omega,b)),\pi(\omega,b)}(d\varphi)\right].$$

Thus we proved (5.24) since by Lemma 3.9 $\tilde{\Delta}(b, s) = \Delta^*(s)$. Equation (5.23) follows from a similar calculation, this time using in (5.27) for any measurable $D \subset \Xi$

$$m_D(\omega, s, t) := \tilde{\Delta}(t, s)\frac{1}{|C(\omega, s)|}\int \mathbf{1}\{\varphi^{-1}C(\omega, s) \in D\}\kappa_{\beta(s),s}(d\varphi).$$

The third equation (5.25) follows from using instead

$$m_D(\omega, s, t) := \int \mathbf{1}\{\varphi^{-1}C(\omega, s) \in D\}\kappa_{\beta(s),s}(d\varphi),$$

while the forth equation (5.26) stems from using

$$m_D(\omega, s, t) := \frac{1}{|C(\omega, s)|}\int \mathbf{1}\{\varphi^{-1}C(\omega, s) \in D\}\kappa_{\beta(s),s}(d\varphi).$$

The assertion about Δ^* is clear in view of Lemma 3.13. We now show the finiteness of all above terms. As O is finite we may conclude from (4.10) that

$$\mathbb{Q}^\xi(\Omega \times O) = \sum_{b \in O} \frac{\mathbb{P}(b \in \xi)}{\lambda(G_{b,b})} < \infty,$$

which shows finiteness of the left side of (5.23). Since by Theorem 4.11

$$\mathbb{P}(\,\cdot\,||\xi)_b = \mathbb{P}(\,\cdot\,|\xi\{b\} = 1), \quad b \in O,$$

it is evident from (5.16) that

$$\int |C(\omega, b)|\mathbb{P}(d\omega||\xi)_b < \infty.$$

Thus from (4.10) the finiteness of the left side of (5.24) follows. The finiteness of the right side of (5.25) is evident just as that of the right side of (5.26). □

We now turn to probabilistic interpretations of these results. Fixing in Γ a (finite) complete system O of orbit representatives we denote by $\mathbb{Q}^\xi$ the cumulative Palm measure of ξ with respect to O. As $0 < \mathbb{Q}^\xi(\Omega \times O) = (\mathbb{E}\xi)^*(O) < \infty$ we may similarly as in Definition 4.23 put

$$\mathbb{P}^\xi := \frac{1}{\mathbb{Q}^\xi(\Omega \times O)}\mathbb{Q}^\xi$$

where we decided to make the dependence on ξ rather than the dependence on $I = \Omega \times O$ (cf. Subsection 4.3.1) explicit. $\mathbb{E}^\xi$ denotes integration with respect to $\mathbb{P}^\xi$. The finiteness of all expressions in the previous theorem allows us to define the following probability measures on Ξ.

Definition 5.11 (various typical clusters). In the situation of Theorem 5.10 we interpret $(\omega, b) \mapsto C(\omega, b)$ as a random subgraph of Γ defined on the space $(\Omega \times O, \mathcal{A} \otimes \mathcal{P}(O), \mathbb{P}^\xi)$. A random subgraph with distribution

(i) $\mathbb{P}^\xi(C \in \cdot)$ is called *typical cluster* of ϑ,

(ii) $\dfrac{1}{\mathbb{E}^\xi|C|} \displaystyle\int \mathbf{1}\{C(\omega, b) \in \cdot\}|C(\omega, b)|\mathbb{P}^\xi(d(\omega, b))$ is called *cluster-size-weighted typical cluster* of ϑ,

(iii) $\dfrac{1}{\mathbb{E}^\xi \sum_{v \in C} \Delta^*(v)} \displaystyle\int \mathbf{1}\{C(\omega, b) \in \cdot\} \sum_{v \in C(\omega, b)} \Delta^*(v)\mathbb{P}^\xi(d(\omega, b))$ is called Δ-*cluster-size-weighted typical cluster* of ϑ.

(iv) $\dfrac{1}{\mathbb{E}^\xi \sum_{v \in C} \lambda(G_{v,v})} \displaystyle\int \mathbf{1}\{C(\omega, b) \in \cdot\} \sum_{v \in C(\omega, b)} \lambda(G_{v,v})\mathbb{P}^\xi(d(\omega, b))$ is called *stabilizer-cluster-size-weighted typical cluster* of ϑ.

Further, let U be uniformly distributed in O and independent of ϑ. Then we call

(v) a random subgraph of Γ with the same distribution as

$$\omega \mapsto C(\omega, U(\omega))$$

a *0-cluster* of ϑ,

(vi) a random subgraph of Γ with distribution

$$\mathbb{E}\left[\frac{1}{|O|}\sum_{b\in O}\int \mathbf{1}\{\varphi^{-1}C(\omega,b)\in\cdot\}\kappa_{\beta(\pi(b)),\pi(b)}(d\varphi)\right]$$

a *centralized 0-cluster* of ϑ,

(vii) a random subgraph of Γ with distribution

$$\mathbb{E}\left[\sum_{b\in O}\frac{1}{S\cdot\lambda(G_{b,b})}\int \mathbf{1}\{\varphi^{-1}C(\omega,b)\in\cdot\}\kappa_{\beta(\pi(b)),\pi(b)}(d\varphi)\right],$$

where $S:=\sum_{b\in O}\frac{1}{\lambda(G_{b,b})}$, a *stabilizer-weighted-picked centralized 0-cluster* of ϑ,

(viii) a random subgraph of Γ with distribution

$$\mathbb{E}\left[\sum_{b\in O}\frac{1}{S\cdot\lambda(G_{b,b})}\Delta^*(\pi(b))\int \mathbf{1}\{\varphi^{-1}C(\omega,b)\in\cdot\}\kappa_{\beta(\pi(b)),\pi(b)}(d\varphi)\right],$$

where $S:=\mathbb{E}\sum_{b\in O}\frac{\Delta^*(\pi(b))}{\lambda(G_{b,b})}$, a *$\Delta$-weighted-picked centralized 0-cluster* of ϑ,

(viii) a random subgraph of Γ with distribution

$$\mathbb{E}\left[\sum_{b\in O}\frac{1}{S\cdot\lambda(G_{b,b})}\frac{\Delta^*(\pi(b))}{|C(b)|}\int \mathbf{1}\{\varphi^{-1}C(\omega,b)\in\cdot\}\kappa_{\beta(\pi(b)),\pi(b)}(d\varphi)\right],$$

where $S:=\mathbb{E}\sum_{b\in O}\frac{\Delta^*(\pi(b))}{\lambda(G_{b,b})|C(b)|}$, a *$\Delta$-weighted-picked centralized size-debiased 0-cluster* of ϑ,

These definitions already indicate the content of the next result, which represents a simple reformulation of Theorem 5.10.

Corollary 5.12 (probabilistic interpretations). *In the situation of Theorem 5.10 let Z denote a typical cluster, Z_s a cluster-size-weighted typical cluster, Z_s^Δ a Δ-cluster-size-weighted typical cluster and Z_s^λ a stabilizer-cluster-size-weighted typical cluster. Let further N denote a centralized 0-cluster, N_w^λ a stabilizer-weighted-picked centralized 0-cluster, N_w^Δ Δ-weighted-picked centralized 0-cluster and n_w^Δ a Δ-weighted-picked centralized size-debiased 0-cluster. Then their distributions are related as follows.*

$$\mathbb{P}(Z\in\cdot)=\mathbb{P}(n_w^\Delta\in\cdot),\tag{5.28}$$

$$\mathbb{P}(Z_s\in\cdot)=\mathbb{P}(N_w^\Delta\in\cdot),\tag{5.29}$$

$$\mathbb{P}(Z_s^\Delta\in\cdot)=\mathbb{P}(N_w^\lambda\in\cdot),\tag{5.30}$$

$$\mathbb{P}(Z_s^\lambda\in\cdot)=\mathbb{P}(N\in\cdot).\tag{5.31}$$

and we have the relations

$$\sum_{b \in O} \frac{\mathbb{P}(b \in \xi)}{\lambda(G_{b,b})} = \mathbb{E}\left[\sum_{b \in O} \frac{1}{\lambda(G_{b,b})} \frac{\Delta^*(\pi(b))}{|C(\pi(b))|}\right], \tag{5.32}$$

$$\mathbb{E}^{\xi}\left[|C|\right] \sum_{b \in O} \frac{\mathbb{P}(b \in \xi)}{\lambda(G_{b,b})} = \mathbb{E}\left[\sum_{b \in O} \frac{\Delta^*(\pi(b))}{\lambda(G_{b,b})}\right], \tag{5.33}$$

$$\mathbb{E}^{\xi}\left[\sum_{v \in C} \Delta^*(v)\right] \sum_{b \in O} \frac{\mathbb{P}(\xi \in O)}{\lambda(G_{b,b})} = \sum_{b \in O} \frac{1}{\lambda(G_{b,b})}, \tag{5.34}$$

$$\mathbb{E}^{\xi}\left[\sum_{v \in C} \lambda(G_{v,v})\right] \frac{1}{|O|} \sum_{b \in O} \frac{\mathbb{P}(b \in \xi)}{\lambda(G_{b,b})} = 1. \tag{5.35}$$

All sides of these equations are finite.

Proof. Rewriting all sides of the equations in Theorem 5.10 in terms of the distributions introduced in Definition 5.11 yields the equations

$$\sum_{b \in O} \frac{\mathbb{P}(b \in \xi)}{\lambda(G_{b,b})} \mathbb{P}(Z \in \cdot) = \mathbb{E}\left[\sum_{b \in O} \frac{1}{\lambda(G_{b,b})} \frac{\Delta^*(\pi(b))}{|C(\pi(b))|}\right] \mathbb{P}(n_w^{\Delta} \in \cdot),$$

$$\mathbb{E}^{\xi}\left[|C|\right] \sum_{b \in O} \frac{\mathbb{P}(b \in \xi)}{\lambda(G_{b,b})} \mathbb{P}(Z_s \in \cdot) = \mathbb{E}\left[\sum_{b \in O} \frac{\Delta^*(\pi(b))}{\lambda(G_{b,b})}\right] \mathbb{P}(N_w^{\Delta} \in \cdot),$$

$$\mathbb{E}^{\xi}\left[\sum_{v \in C} \Delta^*(v)\right] \sum_{b \in O} \frac{\mathbb{P}(\xi \in O)}{\lambda(G_{b,b})} \mathbb{P}(Z_s^{\Delta} \in \cdot) = \sum_{b \in O} \frac{1}{\lambda(G_{b,b})} \mathbb{P}(N_w^{\lambda} \in \cdot),$$

$$\mathbb{E}^{\xi}\left[\sum_{v \in C} \lambda(G_{v,v})\right] \frac{1}{|O|} \sum_{b \in O} \frac{\mathbb{P}(b \in \xi)}{\lambda(G_{b,b})} \mathbb{P}(Z_s^{\lambda} \in \cdot) = \mathbb{P}(N \in \cdot).$$

Plugging in Ξ everywhere then yields the relations (5.32), (5.33), (5.34) and (5.35) and then using these relations with the above equations again finally yields (5.28), (5.29), (5.30) and (5.31). □

5.4.2 Transitive possibly non-unimodular graphs

If the underlying graph Γ is even transitive, then clearly the formulas in Theorem 5.10 and Corollary 5.12 simplify since the summation over O becomes superfluous and 0-clusters then really arise from fixing a vertex and looking at the cluster in which the fixed vertex is contained without any further randomization on how this vertex is chosen. Also, the distributions of the several types of typical clusters defined in Definition 5.11 simplify as one may identify $\Omega \times O = \Omega \times \{o\}$ where $o \in V$ is fixed and arbitrary with Ω. Further examples of transitive non-unimodular graphs (other than $\xi(T_n)$) may be found in [67, 44]. Also, Lemma 3.13 allows explicit expressions for Δ^*. We shall carry this through here for $\xi(T_n)$. Clearly, given two vertices s and t in $\xi(T_n)$ there is a unique youngest common ξ-ancestor of s and t in T_n. More precisely, in T_n, the unique rays $s_{\xi} \in \xi$ and $t_{\xi} \in \xi$ starting in s resp. t, must, since they are equivalent, intersect in a point $\xi(s,t)$ and coalesce behind this point since their remaining parts must coincide with $\xi(s,t)_{\xi}$, the unique ray starting in $\xi(s,t)$ and lying in ξ. The vertex $\xi(s,t)$ is what we called above *youngest common ξ-ancestor*. We shall now compute $\tilde{\Delta}$ for $\xi(T_n)$ as well as Δ^* with respect to an arbitrary fixed vertex o serving as our single orbit representative.

We define

$$L_\xi(v,w) := d(v, \xi(v,w)) - d(w, \xi(v,w)), \quad v,w \in V, \tag{5.36}$$

where $d(v,w)$ denotes the graph distance in T_n (not in $\xi(T_n)$!). Note that $L_\xi(v,w)$ measures the relative ξ-age of v and w. It is reasonable to say that v is *older* than w if $L_\xi(v,w) > 0$, that v is *younger* than w if $L_\xi(v,w) < 0$ and that v and w are *of the same generation* if $L_\xi(v,w) = 0$. Clearly, L_ξ is jointly $\mathrm{Aut}(\xi(T_n))$-invariant.

Lemma 5.13. ($\tilde{\Delta}$ and Δ^* for $\xi(T_n)$) *For* $\mathrm{Aut}(\xi(T_n)) \hookrightarrow V$ *we have*

$$\tilde{\Delta}(v,w) = (n-1)^{L_\xi(v,w)}, \quad v,w \in V, \tag{5.37}$$

and with respect to a fixed vertex o serving as orbit representative

$$\Delta^*(w) = (n-1)^{L_\xi(o,w)}, \quad w \in V. \tag{5.38}$$

Here L_ξ is defined as in (5.36).

Proof. We first compute $\tilde{\Delta}$. Equation (3.15) shows that $\tilde{\Delta}$ does not depend on the choice of our single orbit representative o. Thus we may choose a particular convenient one in Lemma 3.14. For given $v,w \in V$ we choose $O = \{\xi(v,w)\}$ in Lemma 3.14, where $\xi(v,w)$ is the unique youngest common ξ-ancestor of v and w. Thus

$$\tilde{\Delta}(v,w) = \frac{\Delta^*_{\xi(v,w)}(w)}{\Delta^*_{\xi(v,w)}(v)}$$

where $\Delta^*_{\xi(v,w)}$ stands for Δ^* with respect to $O = \{\xi(v,w)\}$. By Lemma 3.13

$$\Delta^*_{\xi(v,w)}(w) = \frac{|G_{w,w}\xi(v,w)|}{|G_{\xi(v,w),\xi(v,w)}w|}.$$

Here $|G_{w,w}\xi(v,w)| = 1$ since any automorphism fixing w must clearly fix any ξ-ancestor of w as well. To evaluate the denominator let k denote the length of the unique path in T_n connecting w and $\xi(v,w)$. There are evidently $(n-1)^k$ ξ-descendants of $\xi(v,w)$ that are exactly k generations younger than $\xi(v,w)$. The set of these descendants equals the set $G_{\xi(v,w),\xi(v,w)}w$, since first, the relation 's is exactly k generations younger than t' is jointly $\mathrm{Aut}(\xi(T_n))$-invariant, and second, any such two descendants may be mapped to each other by means of a $\varphi \in G_{\xi(v,w),\xi(v,w)}$. Thus

$$\Delta^*_{\xi(v,w)}(w) = \frac{1}{(n-1)^k} = \left(\frac{1}{n-1}\right)^{d(w,\xi(v,w))}$$

and since an analogues equality holds for v we receive

$$\tilde{\Delta}(v,w) = \left(\frac{1}{n-1}\right)^{d(w,\xi(v,w)) - d(v,\xi(v,w))}$$

and thus (5.37). Equation (5.38) is now a special case. $\qquad\square$

Figure 5.5 shows some values of Δ^* with respect to a fixed $o \in V$. Clearly Δ^* induces a partial ordering on V where $v \leq w$ if and only if v is younger than or of the same generation as w. This ordering may be used to derive the existence of a center function in this special case, even as a deterministic function of the configuration $\vartheta(\omega)$. Namely, every cluster C of any subgraph of $\xi(T_n)$ has a unique oldest vertex $\pi(C)$, as is easy to see, and clearly $\pi(\varphi C) = \varphi \pi(C)$ and $\pi(C) \in C$. Thus this π satisfies (5.17) and (5.18). Clearly, for $\Gamma = \xi(T_n)$ and with respect to this π and $O = \{o\}$, both the distributions in (5.28), (5.29),(5.30) and (5.31) as well as the formulas (5.32), (5.33),(5.34) and (5.35) become completely explicit by using our formula for Δ^* from (5.38).

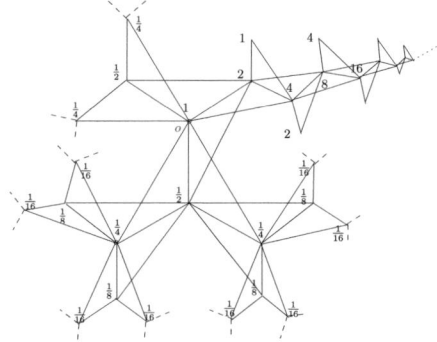

Figure 5.5: Some values of Δ^* with respect to the fixed vertex o.

Chapter 6

Ergodic Theory

For historical information on the development of ergodic theory, we refer to [28, p. 576]. In this chapter we shall not prove a fundamentally new result in ergodic theory, but yet, we shall adapt two results from this field such that they may serve for the inspection of either $\mathbb{Z}^d$-stationary random measures or L-stationary random measures in $\mathbb{R}^d$. More precisely, we consider the actions $\mathbb{Z}^d \hookrightarrow \mathbb{R}^d$ and $L \hookrightarrow \mathbb{R}^d$ where L is a k-dimensional ($0 \leq k < d$) linear subspace of $\mathbb{R}^d$ and both actions are via translation. In either case, if G denotes the respective group, we investigate a G-stationary random measure ξ on $\mathbb{R}^d$ and investigate a.s.- and L^p-convergence of random sequences of the form

$$\frac{\xi(A \cap B_n)}{\lambda^d(A \cap B_n)}, \quad n \in \mathbb{N},$$

where A is a G-invariant set, B_n a sequence of nested increasing G-symmetric sets and λ^d denotes d-dimensional Lebesgue measure on $\mathbb{R}^d$. This will be done by applying two classical (multivariate) ergodic theorems, the first obtained by Zygmund [74] and the second by Wiener [72]. The first case $\mathbb{Z}^d \hookrightarrow \mathbb{R}^d$ will be treated in Section 6.1, while the second case $L \hookrightarrow \mathbb{R}^d$ will be treated in Section 6.2. We fix counting measure as Haar measure on $\mathbb{Z}^d$ and u-dimensional Lebesgue measure λ_L as Haar measure on L. We recall (see the examples in the end of Subsection 2.2.4) that in the case $\mathbb{Z}^d \hookrightarrow \mathbb{R}^d$ a $\mathbb{Z}^d$-invariant set is derived by taking a certain pattern within the half-open unit cube and extending this pattern $\mathbb{Z}^d$-periodically on all of $\mathbb{R}^d$, since the orbits are given by the translates $q + \mathbb{Z}^d, q \in [0,1)^d$. On the other hand $\mathbb{Z}^d$-symmetric subsets of $\mathbb{R}^d$ are non-empty finite unions of the translates $z + [0,1)^d, z \in \mathbb{Z}^d$. Each such set B has the property $\delta(B) \in \mathbb{N}$. In the other case L-invariant subsets of $\mathbb{R}^d$ are unions of translates of L while the prime examples of L-symmetric subsets of $\mathbb{R}^d$ are unions of translates of $L^\perp$ (note that there are many other possibilities to construct L-symmetric subsets of $\mathbb{R}^d$). See Figure 6.1 for illustrations.

Our ergodic theorems will enable us to define analogues of the classical *sample intensity* of a completely stationary random measure on $\mathbb{R}^d$ also for G-stationary random measures, where G is either a finitely generated additive subgroup of $\mathbb{R}^d$ or any proper linear subspace. These analogues are in fact a family of random measures indexed by the σ-algebra of G-invariant sets, and they come out naturally as limits of the above described sequences.

After we derived these ergodic theorems for $\mathbb{Z}^d$-stationary resp. L-stationary random measures on $\mathbb{R}^d$, we show in Section 6.3 how the cumulative Palm measure

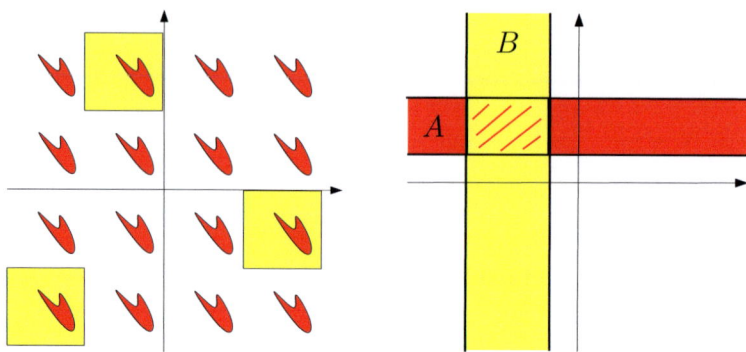

Figure 6.1: Left: $\mathbb{Z}^d \hookrightarrow \mathbb{R}^d$, right: $L \hookrightarrow \mathbb{R}^d$ where $L := \{(x,0) : x \in \mathbb{R}\}$

naturally arises in the limit of certain sequences of the above type under an ergodicity assumption. *Ergodicity* with respect to an operating group may be defined in a completely general framework. Consider a G-stationary random element ξ in a measurable space S, defined on an underlying probability space $(\Omega, \mathcal{A}, \mathbb{P})$, where the group G acts on S in some way. As before, we may assume without loss of generality the existence of a measurable flow θ indexed by G on Ω and such that

$$\xi(\theta_g \omega) = g\xi(\omega), \quad \omega \in \Omega, g \in G.$$

The symbol $\mathcal{I}$ denotes the σ-algebra of G-invariant measurable subsets of S and we put

$$\mathcal{I}_\xi = \{\{\xi \in A\} : A \in \mathcal{I}\}.$$

Now ξ is called *G-ergodic*, if $\mathcal{I}_\xi$ is $\mathbb{P}$-trivial, i.e.

$$\mathbb{P}(\xi \in A) \in \{0,1\}, \quad A \in \mathcal{I}.$$

The importance of this notion will become clear when we state the two classical ergodic theorems announced above. $\mathcal{B}^d$ denotes the Borel σ-algebra in $\mathbb{R}^d$.

6.1 Grid-stationary random measures

We treat the case of $\mathbb{Z}^d$-stationary random measures first, since this is easier to handle than the L-stationary case.

6.1.1 An ergodic theorem for lattice-actions

The aim of this subsection is to provide the necessary tool needed to derive an ergodic theorem for $\mathbb{Z}^d$-stationary random measures on $\mathbb{R}^d$. Given a measure μ on a space S, and a map $T : S \to S$, we call T μ-preserving, or, changing perspective, μ T-invariant, if

$$\mu \circ T^{-1} = \mu.$$

We further define, given a measure space $(S, \mathcal{S}, \mu)$, the classes

$$L \log^m L(\mu) := \left\{ f \in \mathcal{S}_+ : \int |f(s)| \log_+^m(|f(s)|)\mu(ds) < \infty \right\}, \quad m \geq 0,$$

where for a given function $g : \mathbb{R} \to \mathbb{R}$ its *positive part* g_+ is defined as $x \mapsto \max\{g(x), 0\}$. Given a σ-algebra $\mathcal{J}$ on Ω, we write $\mathbb{E}^{\mathcal{J}}[\cdot] := \mathbb{E}[\cdot|\mathcal{J}]$. The following theorem, literally taken from [28, Theorem 10.12], has been derived by Zygmund [74]. It represents a discrete multivariate ergodic theorem even for possibly non-commuting transformations.

Theorem 6.1 (Zygmund's multivariate ergodic theorem). *Let ξ denote a random element in a space S with distribution μ, $T_1, \ldots, T_d$ be μ-preserving maps on S with invariant σ-fields $\mathcal{I}_1, \ldots, \mathcal{I}_d$ and put $\mathcal{J}_k := \xi^{-1}\mathcal{I}_k$. Then for any $f \in L \log^{d-1} L(\mu)$ we have as $n_1, \ldots, n_d \to \infty$*

$$\frac{1}{n_1 \cdot \ldots \cdot n_d} \sum_{0 \leq k_1 < n_1} \cdots \sum_{0 \leq k_d < n_d} f(T_1^{k_1} \ldots T_d^{k_d}\xi) \to \mathbb{E}^{\mathcal{J}_d} \ldots \mathbb{E}^{\mathcal{J}_1} f(\xi), \quad a.s.$$

The convergence holds in L^p for a $p \geq 1$, whenever $f \in L^p(\mu)$.

We now slightly modify this theorem such that it applies to $\mathbb{Z}^d$-actions. To this end consider an operation $\mathbb{Z}^d \hookrightarrow S$ and denote the associated shift operators by $\theta_z, z \in \mathbb{Z}^d$. We define a *box* B in $\mathbb{Z}^d$ as a set of the form

$$B := (-k_1, k_1] \times \cdots \times (-k_d, k_d] \cap \mathbb{Z}^d, \quad k_i \in \mathbb{N}, i \in \{1, \ldots, d\},$$

and note that a box of this form has $2^d \cdot k_1 \cdot \ldots \cdot k_d$ elements. An *increasing* sequence of boxes B_n is a sequence of boxes where for all $i \in \{1, \ldots, d\}$ we have $k_i(n) \uparrow \infty$ for $n \to \infty$. Then Zygmund's ergodic theorem may be used to prove:

Theorem 6.2 (ergodic theorem for $\mathbb{Z}^d$-actions). *Let ξ denote a $\mathbb{Z}^d$-stationary random element in a space S with distribution μ and B_n an increasing sequence of boxes. Then for any $f \in L \log^{d-1} L(\mu)$ it holds for $n \to \infty$*

$$\frac{1}{|B_n|} \sum_{z \in B_n} f(\theta_z \xi) \to \mathbb{E}[f(\xi)|\mathcal{I}_\xi] \quad a.s.$$

where $\mathcal{I}_\xi := \xi^{-1}\mathcal{I}$, $\mathcal{I}$ denoting the σ-algebra of $\mathbb{Z}^d$-invariant measurable sets in S. The same convergence holds in L^p for a $p \geq 1$ whenever $f \in L^p(\mu)$.

Proof. Denote by e_k the k-th standard unit vector in $\mathbb{Z}^d$ and let T_k denote the shift on S induced by e_k. Note that these T_k are invertible and hence T_k^n makes sense for all $n \in \mathbb{Z}$. By the commutativity of $\mathbb{Z}^d$ there are unique $k_1(z), \ldots, k_d(z)$ for each $z \in \mathbb{Z}^d$ such that $\theta_z = T_1^{k_1(z)} \circ \ldots \circ T_d^{k_d(z)}$, where θ_z denotes the shift on S induced by $z \in \mathbb{Z}^d$. Also, since the T_k commute we have, writing $\mathcal{I}_k$ for the T_k-invariant σ-algebra on S and $\mathcal{J}_k := \xi^{-1}\mathcal{I}_k$,

$$\mathbb{E}^{\mathcal{J}_d} \ldots \mathbb{E}^{\mathcal{J}_1} f(\xi) = \mathbb{E}[f(\xi)|\mathcal{I}_\xi]$$

by [28, Corollary 10.13], since evidently $\bigcap_k \mathcal{J}_k = \xi^{-1} \bigcap_k \mathcal{I}_k = \xi^{-1}\mathcal{I} = \mathcal{I}_\xi$. Any d-dimensional orthant of $\mathbb{R}^d$ will in the following be interpreted as a product of length d and of the factors $(-\infty, 0]$ and $(0, \infty)$ exclusively to ensure that they are disjoint. Label each of the 2^d d-dimensional orthants by the unique d-tuple $(a_1, \ldots, a_d)$, where $a_i \in \{-1, 1\}$, lying inside of it. It remains to split each box into the 2^d different sections with the disjoint orthants $Q_1, \ldots, Q_{2^d}$. In the orthant labeled with $(a_1, \ldots, a_d)$ we may apply Zygmund's multivariate ergodic theorem to the transformations $T_1^{a_1}, \ldots, T_d^{a_d}$ and the random element $T_1^{(a_1+1)/2} \circ \ldots \circ T_d^{(a_d+1)/2} \xi$ which yields

$$\frac{1}{|B_n \cap Q_i|} \sum_{z \in B_n \cap Q_i} f(\theta_z \xi) \to \mathbb{E}[f(\xi)|\mathcal{I}_\xi] \quad a.s. \quad i \in \{1, \ldots, 2^d\},$$

and in L^p under the respective condition. Here we also used the obvious fact that

$$\mathbb{E}\left[f\left(\theta_z \xi\right) \Big| \mathcal{I}_\xi\right] = \mathbb{E}[f(\xi)|\mathcal{I}_\xi] \quad a.s., \quad z \in \mathbb{Z}^d.$$

Finally, since

$$|B_n \cap Q_i|/|B_n| = 1/2^d, \quad i \in \{1, \ldots, 2^d\},$$

we may proceed via

$$\frac{1}{|B_n|} \sum_{z \in B_n} f(\theta_z \xi) = \sum_{i=1}^{2^d} \frac{|B_n \cap Q_i|}{|B_n|} \frac{1}{|B_n \cap Q_i|} \sum_{z \in B_n \cap Q_i} f(\theta_z \xi)$$

$$= \frac{1}{2^d} \sum_{i=1}^{2^d} \frac{1}{|B_n \cap Q_i|} \sum_{z \in B_n \cap Q_i} f(\theta_z \xi)$$

and the assertion follows. □

6.1.2 Sample intensity for grid-stationary random measures

Classically, *sample intensities* of random measures have been defined for 'completely' stationary random measures such as random measures on groups stationary with respect to the canonical action of the group on itself via left-translations or on homogeneous spaces. Here we show how to introduce such an object for $\mathbb{Z}^d$-stationary random measures on $\mathbb{R}^d$. As it turns out, the relevant object will be a collection of random variables, indexed by the collection of $\mathbb{Z}^d$-invariant measurable subsets A of $\mathbb{R}^d$ that satisfy the reasonable condition that $\lambda^d(A \cap [0, 1)^d) > 0$. We call such $\mathbb{Z}^d$-invariant sets *admissible*. As usual λ^d denotes d-dimensional Lebesgue measure on $\mathbb{R}^d$ and we define

$$\bar{\xi}_A := \frac{\mathbb{E}[\xi(A \cap [0, 1)^d)|\mathcal{I}_\xi]}{\lambda^d(A \cap [0, 1)^d)},$$

for any admissible A. The following theorem shows that we may interpret the quantity $\bar{\xi}_A(\omega)$ as the intensity of the sample $\xi(\omega)$ on A.

Theorem 6.3 (sample intensity for $\mathbb{Z}^d$-stationary random measures). *Let ξ denote a $\mathbb{Z}^d$-stationary random measure on $\mathbb{R}^d$, let A be a $\mathbb{Z}^d$-invariant admissible measurable subset of $\mathbb{R}^d$ and B_n a sequence of $\mathbb{Z}^d$-symmetric subsets of $\mathbb{R}^d$ such that*

$B_n \cap \mathbb{Z}^d$ *is an increasing sequence of boxes. Then, if* $\xi([0,1)^d \cap A) \in L \log^{d-1} L(\mathbb{P})$, *we have*

$$\frac{\xi(A \cap B_n)}{\lambda^d(A \cap B_n)} \to \bar{\xi}_A, \quad a.s.$$

The same convergence holds in L^p *for some* $p \geq 1$ *whenever* $\xi(A \cap [0,1)^d) \in L^p$.

Proof. Consider the function $f : \mathbf{M}(S) \to \mathbb{R} \cup \{\infty\}, f(\nu) := \nu([0,1)^d \cap A)$. Since by $\mathbb{Z}^d$-invariance of A

$$\lambda(A \cap B_n) = |B_n \cap \mathbb{Z}^d| \lambda(A \cap [0,1)^d),$$

some manipulation yields

$$\frac{\xi(A \cap B_n)}{\lambda(A \cap B_n)} = \frac{1}{\lambda(A \cap [0,1)^d)} \frac{1}{|B_n \cap \mathbb{Z}^d|} \sum_{z \in B_n \cap \mathbb{Z}^d} \xi((z + [0,1)^d) \cap A)$$

$$= \frac{1}{\lambda(A \cap [0,1)^d)} \frac{1}{|B_n \cap \mathbb{Z}^d|} \sum_{z \in B_n \cap \mathbb{Z}^d} \xi(\theta_z^{-1}, [0,1)^d \cap A)$$

$$= \frac{1}{\lambda(A \cap [0,1)^d)} \frac{1}{|B_n \cap \mathbb{Z}^d|} \sum_{z \in B_n \cap \mathbb{Z}^d} f(\theta_z^{-1}\xi).$$

Here we may apply Theorem 6.2 to the operation $\tilde{\theta}_z\mu := \theta_z^{-1}\mu, z \in \mathbb{Z}^d$, and since the associated invariant σ-algebra $\tilde{\mathcal{I}}$ clearly coincides with $\mathcal{I}$, i.e. $\tilde{\mathcal{I}}_\xi = \mathcal{I}_\xi$, this yields the respective assertions. $\qquad\square$

Remark 6.4. We note that $\bar{\xi}_A(\omega)$ is far from being a measure in A - it is not even finitely additive. Also, the method of the above proof, namely to consider the induced $\mathbb{Z}^d$-stationary random measure $\eta(\{z\}) := \xi((z + [0,1)^d) \cap A), z \in \mathbb{Z}^d$, on $\mathbb{Z}^d$ raises the question if $\mathbb{Z}^d$-ergodicity of ξ implies that of η. This is true: since $\eta = f(\xi)$ with $\mathbb{Z}^d$-covariant (!) $f : \mathbf{M}(S) \to \mathbf{M}(S)$ it follows that $\mathcal{I}_\eta \subset \mathcal{I}_\xi$ and hence $\mathbb{P}$-triviality of $\mathcal{I}_\xi$ implies that of $\mathcal{I}_\eta$.

Remark 6.5. Theorem 6.3 sheds light on the notion of G-ergodicity of a random measure, as defined in the end of the introduction to this chapter. It says that

$$\frac{\xi(A \cap B_n)}{\lambda^d(A \cap B_n)} \to \bar{\xi}_A$$

either a.s. or in L^p under respective mild conditions on $\xi(A \cap [0,1)^d)$. Now if ξ is $\mathbb{Z}^d$-ergodic, then this limit equals by $\mathbb{P}$-triviality of $\mathcal{I}_\xi$

$$\bar{\xi}_A = \frac{\mathbb{E}[\xi(A \cap [0,1)^d)]}{\lambda^d(A \cap [0,1)^d)} \quad a.s.$$

and is thus constant. The important message is the following intuition about G-ergodicity. While G-stationarity enforces a spatial homogeneity of the random measure along each orbit, G-ergodicity enforces a uniformity in $\omega \in \Omega$ *on every single orbit* (sometimes Ω is in this context also called *phase space* with the intuition that an ergodic process is always in the same phase or modus, while the phases of non-ergodic processes may change). The above theorem shows that $\mathbb{Z}^d$-ergodicity must show simultaneously on every single fixed union of $\mathbb{Z}^d$-orbits. The same intuition will apply to ergodicity with respect to an operating linear subspace in the next section.

6.2 Partly stationary random measures

In this section, we consider L-stationary random measures on $\mathbb{R}^d$, where L is a fixed linear subspace of $\mathbb{R}^d$ that acts on $\mathbb{R}^d$ via translation. The induced shifts on $\mathbf{M}(\mathbb{R}^d)$ and the abstract flow on Ω are both (abusing notation) denoted by θ. The appropriate ergodic theorem that we will work with will be stated in the following Subsection 6.2.1 along with some further lemmas that we will need later. We then derive a result similar to that in Theorem 6.3 in Subsection 6.2.2.

6.2.1 Wiener's ergodic theorem and further preparations

For convex sets $B \subset \mathbb{R}^d$ we denote by $r(B)$ the *inner radius* of B, i.e. the radius of the largest open ball contained in B. We recall the following classical *spatial* (also called *multivariate*) *ergodic theorem* by Wiener [72]. A convenient reference for a streamlined and thoroughly worked out proof is again [28, Theorem 10.14].

Theorem 6.6 (spatial ergodic theorem, Wiener). *Let ξ be a random element in a measurable space S with distribution μ and assume that μ is $\mathbb{R}^d$-invariant, i.e. θ-invariant. Fix some bounded, convex measurable sets $B_1 \subset B_2 \subset \ldots$ with $r(B_n) \to \infty$. Then for any $f \in \mathcal{S}_+$*

$$\frac{1}{\lambda^d(B_n)} \int_{B_n} f(\theta_s \xi) \lambda^d(ds) \to \mathbb{E}[f(\xi)|\mathcal{I}_\xi] \quad a.s.$$

If $f \in L^p(\mu)$ for some $p \geq 1$ then the same convergence holds in $L^p(\mu)$.

Nguyen and Zessin [57] proved the following result on sample intensities of completely stationary random measures in $\mathbb{R}^d$. It may be derived as a consequence of Theorem 6.6, cf. [28, Corollary 10.19].

Theorem 6.7 (completely stationary case, Nguyen, Zessin). *Let ξ be a stationary random measure on $\mathbb{R}^d$ and fix some bounded convex sets $B_1 \subset B_2 \subset \ldots$ with $r(B_n) \to \infty$. Then*

$$\frac{\xi(B_n)}{\lambda^d(B_n)} \to \bar{\xi} \quad a.s.$$

where for some fixed $C \in \mathcal{B}^d$ with $0 < \lambda^d(C) < \infty$

$$\bar{\xi} := \frac{\mathbb{E}[\xi(C)|\mathcal{I}_\xi]}{\lambda^d(C)}.$$

The same convergence also holds in L^p for some $p \geq 1$ whenever $\xi([0,1]^d) \in L^p$.

This also implies that $\bar{\xi}$ is well-defined indeed, i.e. does not depend on C a.s. since the approximating sequence is independent of C. We shall also need the following result on convex sets, taken from [28, Lemma 10.15 (ii)] (but stated there without proof, which is why we provide one here). For a set $K \subset \mathbb{R}^d$ and $\varepsilon > 0$ let $\partial_\varepsilon K$ denote the ε-neighborhood of ∂K, and B^d the open unit ball in $\mathbb{R}^d$.

Lemma 6.8 (convex sets). *If $B \subset \mathbb{R}^d$ is convex and bounded with $r(B) > 0$, then for any $\varepsilon > 0$*

$$\lambda^d(\partial_\varepsilon B) \leq 2 \left(\left(1 + \frac{\varepsilon}{r(B)} \right)^d - 1 \right) \lambda^d(B).$$

Proof. Fix a convex set B and $\varepsilon > 0$ and put $\partial_\varepsilon^+ B := \partial_\varepsilon B \cap B^c$ and $\partial_\varepsilon^- B := \partial_\varepsilon B \cap B$. Then $\partial_\varepsilon B = \partial_\varepsilon^+ B \cup \partial_\varepsilon^- B$ and since B is convex we have

$$\partial_\varepsilon^+ B = (B + \varepsilon B^d) \setminus B.$$

It is enough to show that

$$\lambda^d(\partial_\varepsilon^\pm B) \leq \left(\left(1 + \frac{\varepsilon}{r(B)} \right)^d - 1 \right) \lambda^d(B).$$

The assertion for $\partial_\varepsilon^+ B$ may be seen as follows. Take an open ball contained in B with center x and radius ρ. Then since $x + \rho B^d \subset B$ it follows that $\varepsilon B^d \subset \frac{\varepsilon}{\rho}(B - x)$. Thus

$$\lambda^d(\partial_\varepsilon^+ B) = \lambda^d(B + \varepsilon B^d) - \lambda^d(B) \leq \lambda^d \left(B + \frac{\varepsilon}{\rho}(B - x) \right) - \lambda^d(B)$$

$$= \lambda^d \left(\left(1 + \frac{\varepsilon}{\rho} \right) B - \frac{\varepsilon}{\rho} x \right) - \lambda^d(B) = \left(1 + \frac{\varepsilon}{\rho} \right)^d \lambda^d(B) - \lambda^d(B)$$

and we are done. The respective assertion for $\partial_\varepsilon^- B$ then follows from the inequality

$$\lambda^d(\partial_\varepsilon^- B) \leq \lambda^d(\partial_\varepsilon^+ B). \tag{6.1}$$

To prove (6.1), take a 1-Lipschitz measurable map

$$F : \partial_\varepsilon^+ B \to \mathbb{R}^d$$

with the property that $\partial_\varepsilon^- B \subset F(\partial_\varepsilon^+ B)$. The existence of such a map will be insured by the next Lemma 6.9. Since 1-Lipschitz maps cannot increase Lebesgue measure (which is most easily seen by invoking the well-known equality of Lebesgue measure with d-dimensional Hausdorff measure up to constant, the advantage being here that the latter is defined in terms of diameters where the 1-Lipschitz property may be applied directly) we have

$$\lambda^d(\partial_\varepsilon^+ B) \geq \lambda^d(F(\partial_\varepsilon^+ B)) \geq \lambda^d(\partial_\varepsilon^- B),$$

and we are done.   □

In the above proof we used the following lemma. The *metric projection* $p(A, x)$ of a point $x \in \mathbb{R}^d$ on a closed convex subset A of $\mathbb{R}^d$ is defined as the unique closest point in A to x.

Lemma 6.9 (existence of 1-Lipschitz map with out-in-property). *Given a convex set B and $\varepsilon > 0$ the map*

$$F : \partial_\varepsilon^+ B \to \mathbb{R}^d, \quad x \mapsto x + 2(p(\bar{B}, x) - x) = 2p(\bar{B}, x) - x,$$

where $p(\bar{B}, x)$ denotes the metric projection of $x \in \mathbb{R}^d$ on the closure of B, satisfies

$$\partial_\varepsilon^- B \subset F(\partial_\varepsilon^+ B)$$

and is 1-Lipschitz.

Proof. To prove the inclusion take $x \in \partial_\varepsilon^- B = \partial_\varepsilon B \cap B$. Let $I(x)$ denote the open ball with center x and radius $d(x, \partial B) \leq \varepsilon$. Choose any $y \in \partial B \cap \partial I(x)$ and put $z := 2y - x$. Then $z = y + (y - x) \in \partial_\varepsilon^+ B$ since $y \in \partial B$, $||y - x|| \leq \varepsilon$ and $z \in B^c$ since B is convex. Clearly $p(\bar{B}, z) = y$ and hence

$$F(z) = 2p(\bar{B}, z) - z = 2y - z = x.$$

To prove the 1-Lipschitz continuity fix $x, y \in \mathbb{R}^d$. Then since the metric projection onto convex sets itself has this property (cf. [62, Theorem 1.2.2]) we have

$$||p(\bar{B}, x) - p(\bar{B}, y)|| \leq ||x - y||.$$

We put $a := y - p(\bar{B}, y)$, $b := x - p(\bar{B}, x)$ and $c := p(\bar{B}, y) - p(\bar{B}, x)$. Then by a simple geometric consideration involving the two hyperplanes with common normal vector $p(\bar{B}, x) - p(\bar{B}, y)$ through $p(\bar{B}, x)$ resp. $p(\bar{B}, y)$ and the above inequality we find that

$$\langle a, c \rangle \geq 0 \quad \text{and} \quad \langle b, c \rangle \leq 0. \tag{6.2}$$

Then since

$$\begin{aligned}
||F(x) - F(y)||^2 &= ||y - x + 2(p(\bar{B}, x) - p(\bar{B}, y))||^2 = ||a - b - c||^2 \\
&= \langle a - b + c - 2c, a - b + c - 2c \rangle \\
&= \langle a - b + c, a - b + c \rangle - 4 \langle a - b + c, c \rangle + 4||c||^2 \\
&= ||x - y||^2 + 4 \left(||c||^2 - \langle a - b + c, c \rangle \right)
\end{aligned}$$

it remains to show that $||c||^2 \leq \langle a - b + c, c \rangle$. But this means nothing but $0 \leq \langle a - b, c \rangle$ which holds by (6.2). $\qquad\square$

Lemma 6.8 leads to the following lemma on special sequences of L-symmetric convex sets. For any convex set B and $a \geq 0$ define $B^+(a) := B + aB^d$ and $B^-(a) := (B^c + aB^d)^c$.

Lemma 6.10 (vanishing thickening/thinning in the limit). *Let $B_1 \subset B_2 \subset \ldots$ denote a nested sequence of convex L-symmetric and $L^\perp$-invariant subsets of $\mathbb{R}^d$ with $\delta(B_n) \to \infty$. Then for any fixed $a \geq 0$*

$$\frac{\delta(B_n^\pm(a))}{\delta(B_n)} \to 1, \quad n \to \infty.$$

Proof. Fix $a > 0$ and let B denote an arbitrary bounded convex set first. Then on one side

$$\frac{\lambda^d(B) - \lambda^d(\partial_a B)}{\lambda^d(B)} \leq \frac{\lambda^d(B) - \lambda^d(\partial_a^- B)}{\lambda^d(B)} = \frac{\lambda^d(B \setminus \partial_a^- B)}{\lambda^d(B)} = \frac{\lambda^d(B^-(a))}{\lambda^d(B)} \leq 1$$

and on the other

$$1 \leq \frac{\lambda^d(B^+(a))}{\lambda^d(B)} \leq \frac{\lambda^d(B \cup \partial_a B)}{\lambda^d(B)} \leq \frac{\lambda^d(B) + \lambda^d(\partial_a B)}{\lambda^d(B)}.$$

Hence Lemma 6.8 yields that for fixed $a \geq 0$

$$\frac{\lambda^d(B^\pm(a))}{\lambda^d(B)} \to 1, \quad r(B) \to \infty. \tag{6.3}$$

If B is in addition L-symmetric and $L^\perp$-invariant then it is easy to see that $B^\pm$ is again L-symmetric and since

$$\delta(B^\pm) = \lambda_L(L \cap B^\pm)$$

we may apply the convergence in (6.3) to the space L which yields the lemma. $\square$

We shall also need the following monotonicity properties of convolutions with respect to thinning and thickening. We recall here that given two functions $f, g : \mathbb{R}^d \to [0, \infty]$, their *convolution* is defined as a new function from $\mathbb{R}^d$ to $[0, \infty]$ via

$$(f * g)(s) = \int f(s - t)g(t)\lambda^d(dt) = \int f(t)g(s - t)\lambda^d(dt), \quad s \in \mathbb{R}^d.$$

Lemma 6.11 (thickening and thinning in convolutions). *For any measurable $C, D \subset \mathbb{R}^d$ with $C \subset aB^d$ for some $a > 0$, we have*

$$\mathbf{1}_C * \mathbf{1}_{D^-(a)} \leq \lambda^d(C)\mathbf{1}_D \leq \mathbf{1}_C * \mathbf{1}_{D^+(a)}. \tag{6.4}$$

Proof. First, we prove the left inequality. For all $x \in \mathbb{R}^d$

$$\mathbf{1}_C * \mathbf{1}_{D^-(a)}(x) = \int \mathbf{1}_C(s)\mathbf{1}_{D^-(a)}(x - s)\lambda^d(ds) \leq \int \mathbf{1}_C(s)\lambda^d(ds) = \lambda^d(C).$$

Further, if $x \notin D$, i.e. if $x \in D^c$, then $C \subset x - (D^c + aB^d) = (x - D^-)^c$ and hence $C \cap (x - D^-) = \emptyset$. This implies

$$\mathbf{1}_C * \mathbf{1}_{D^-(a)}(x) = \int \mathbf{1}_C(s)\mathbf{1}_{x-D^-(a)}(s)\lambda^d(ds) = 0.$$

To prove the right inequality take $x \in D$. Then $C \subset aB^d \subset x - (D + aB^d) = x - D^+(a)$ and

$$\mathbf{1}_C * \mathbf{1}_{D^+(a)}(x) = \int \mathbf{1}_C(s)\mathbf{1}_{x-D^+(a)}(s)\lambda^d(ds) = \int \mathbf{1}_C(s)\lambda^d(ds) = \lambda^d(C).$$

The case $x \notin D$ is trivial. $\square$

In addition, we note the following exchangeability property involving convolutions of subsets of $\mathbb{R}^d$ with symmetry properties (also see Figure 6.2 for a 'proof by a picture' in dimension 2).

Lemma 6.12 (exchangeability property). *Let $A \subset \mathbb{R}^d$ be L-invariant and $B, C \subset \mathbb{R}^d$ both L-symmetric and $L^\perp$-invariant. Then*

$$\mathbf{1}_{A \cap B} * \mathbf{1}_C = \mathbf{1}_B * \mathbf{1}_{A \cap C}. \tag{6.5}$$

Proof. The orbital decomposition of λ^d for $O = L^\perp$ yields for fixed $x \in \mathbb{R}^d$

$$\mathbf{1}_{A \cap B} * \mathbf{1}_C(x) = \int \mathbf{1}_{A \cap B}(s)\mathbf{1}_C(x - s)\lambda(ds) = \iint \mathbf{1}_{A \cap B}(s)\mathbf{1}_C(x - s)\mu_b(ds)\lambda^*(db)$$

$$= \iint \mathbf{1}_A(g + b)\mathbf{1}_B(g + b)\mathbf{1}_C(x - g - b)\lambda_L(dg)\lambda^*(db).$$

Since A is L-invariant and both B and C are $O = L^\perp$-invariant, we have

$$\mathbf{1}_{A \cap B} * \mathbf{1}_C(x) = \iint \mathbf{1}_A(b)\mathbf{1}_B(g)\mathbf{1}_C(x - g)\lambda_L(dg)\lambda^*(db).$$

Now there are unique $x_L \in L$ and $x^\perp \in L^\perp$ such that $x = x_L + x^\perp$. Hence, again by $L^\perp$-invariance of C

$$\mathbf{1}_{A \cap B} * \mathbf{1}_C(x) = \int \mathbf{1}_A(b) \int \mathbf{1}_B(g)\mathbf{1}_C(x_L - g)\lambda_L(dg)\lambda^*(db) = \int \mathbf{1}_A(b)(\mathbf{1}_B *_L \mathbf{1}_C)(x_L)\lambda^*(db),$$

where $*_L$ denotes convolution in L. By the commutativity property of a convolution, the right-hand side is invariant with respect to interchanging B and C. Hence this must also be true for the left-hand side, which gives the assertion. $\qquad\square$

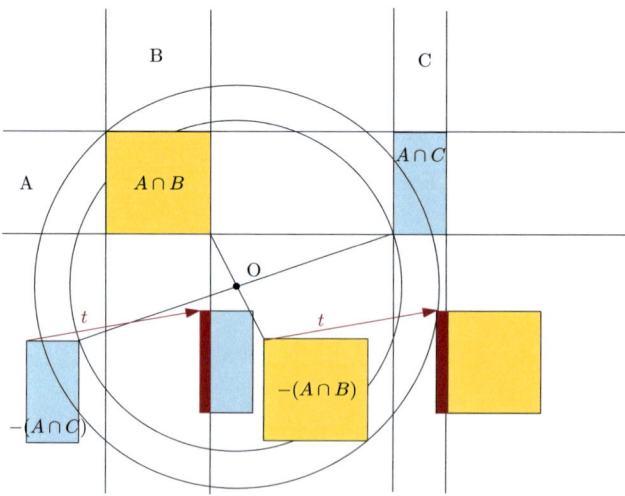

Figure 6.2: A geometric proof of $\mathbf{1}_{A \cap B} * \mathbf{1}_C = \mathbf{1}_B * \mathbf{1}_{A \cap C}$ in dimension 2.

6.2.2 Sample intensity

In the following, we will extend Theorem 6.7 to the case of L-stationary random measures on $\mathbb{R}^d$, where L is a k-dimensional subspace of $\mathbb{R}^d$. Let λ_L denote Haar measure on L, normalized such that a k-dimensional unit cube has measure 1, i.e. k-dimensional Lebesgue measure on L. We shall interpret this measure as a measure on all of $\mathbb{R}^d$ by putting $\lambda_L(\mathbb{R}^d \setminus L) := 0$. As a measurable system of representatives of the orbits, we choose $O := L^{\perp}$, the advantage being here that $O = L^{\perp} \subset \mathbb{R}^d$ is also a group that acts on $\mathbb{R}^d$ in the natural way. As usual, if ν is a σ-finite L-invariant measure on $\mathbb{R}^d$, ν^* denotes the unique measure concentrated on $O = L^{\perp}$ satisfying

$$\int f(x)\nu(dx) = \iint f(x)\mu_b(dx)\nu^*(db), \quad f \in \mathcal{B}_+^d.$$

For fixed $b \in O = L^{\perp}$, the measure $\mu_b = \lambda_L \circ \pi_b^{-1}$ is k-dimensional Hausdorff measure on the affine k-flat $L + b$. Further, for any measure μ on $\mathbb{R}^d$ and $U \in \mathcal{B}^d$ we write

$$\mu_U := \mu(U \cap \cdot)$$

for the restriction of μ to the set U. Analogously to the $\mathbb{Z}^d$-stationary case we now define the sample intensity of ξ on admissible L-invariant subsets A of $\mathbb{R}^d$. Here *admissible* means that there is an L-symmetric and $L^{\perp}$-invariant set B such that $A \cap B$ is bounded and $\lambda^d(A \cap B) > 0$. In this case, changing perspective, we say that B is A-*regular* and we may define

$$\bar{\xi}_A(\omega) := \frac{\mathbb{E}[\xi(A \cap B)|\mathcal{I}_\xi](\omega)}{\lambda^d(A \cap B)}, \quad \omega \in \Omega, A \in \mathcal{I}.$$

The following announced generalization of Theorem 6.7 shows that this quantity does not depend on the choice of B, which justifies as in the $\mathbb{Z}^d$-stationary case our notation. We denote by $B(0, a)$ the open ball around the origin with radius $a \geq 0$.

Theorem 6.13 (sample intensity). *Let L denote a fixed k-dimensional linear subspace of $\mathbb{R}^d$, where $1 \leq k \leq d$. Let further ξ denote an L-stationary random measure on $\mathbb{R}^d$, $A \subset \mathbb{R}^d$ an admissible L-invariant set and $B_1 \subset B_2 \subset \ldots$ a nested sequence of convex L-symmetric and $L^{\perp}$-invariant sets in $\mathbb{R}^d$ with $\delta(B_n) \to \infty$. Then*

$$\frac{\xi(A \cap B_n)}{\lambda^d(A \cap B_n)} \to \bar{\xi}_A, \quad a.s.$$

The same convergence also holds in L^p for given $p \geq 1$ whenever $\xi(A \cap B) \in L^p$ for at least one A-regular set B.

Proof. We modify the proof given by Kallenberg in [28, Corollary 10.19]. Let B denote a fixed A-regular subset of $\mathbb{R}^d$. Fix $a > 0$ such that $A \cap B \subset B(0, a)$ and for any $C \subset \mathbb{R}^d$ put $C^+ := C + B(0, a)$ and $C^- := (C^c + B(0, a))^c$. Then by (6.4)

$$\mathbf{1}_{A \cap B} * \mathbf{1}_{B_n^-} \leq \lambda^d(A \cap B)\mathbf{1}_{B_n} \leq \mathbf{1}_{A \cap B} * \mathbf{1}_{B_n^+}, \quad n \in \mathbb{N}.$$

and hence (recall that $\xi_A := \xi(A \cap \cdot)$), it follows that

$$\frac{\lambda(A \cap B_n^-)}{\lambda(A \cap B_n)} \frac{\xi_A(\mathbf{1}_{A \cap B} * \mathbf{1}_{B_n^-})}{\lambda(A \cap B_n^-)} \leq \lambda(A \cap B) \frac{\xi_A(B_n)}{\lambda(A \cap B_n)} \leq \frac{\lambda(A \cap B_n^+)}{\lambda(A \cap B_n)} \frac{\xi_A(\mathbf{1}_{A \cap B} * \mathbf{1}_{B_n^+})}{\lambda(A \cap B_n^+)}.$$

$$(6.6)$$

Here, using (2.13) and Lemma 6.10

$$\frac{\lambda^d(A \cap B_n^\pm)}{\lambda^d(A \cap B_n)} = \frac{\delta(B_n^\pm)}{\delta(B_n)} \to 1, \quad n \to \infty. \tag{6.7}$$

Fubini's Theorem implies for any $C, D \in \mathcal{B}^d$

$$\int_D (\theta_s^{-1}\xi_A)(C)\lambda(ds) = \iint \mathbf{1}_C(t-s)\mathbf{1}_D(s)\lambda(ds)\xi_A(dt) = \xi_A(\mathbf{1}_C * \mathbf{1}_D). \tag{6.8}$$

By (6.5), (2.13) and (6.8)

$$\frac{\xi_A(\mathbf{1}_{A \cap B} * \mathbf{1}_{B_n^\pm})}{\lambda(A \cap B_n^\pm)} = \frac{\xi_A(\mathbf{1}_B * \mathbf{1}_{A \cap B_n^\pm})}{\lambda(A \cap B_n^\pm)} = \frac{1}{\lambda^*(A)\delta(B_n^\pm)} \int_{A \cap B_n^\pm} (\theta_s^{-1}\xi_A)(B)\lambda(ds). \tag{6.9}$$

Here, using the orbital decomposition $\lambda(ds) = \mu_b(ds)\lambda^*(db)$, we obtain

$$\int_{A \cap B_n^\pm} (\theta_s^{-1}\xi_A)(B)\lambda(ds) = \iint \mathbf{1}\{s \in A \cap B_n^\pm\}(\theta_s^{-1}\xi_A)(B)\mu_b(ds)\lambda^*(db)$$

$$= \iint \mathbf{1}\{g + b \in A, g + b \in B_n^\pm\}(\theta_b^{-1}\theta_g^{-1}\xi_A)(B)\lambda_L(dg)\lambda^*(db)$$

$$= \iint \mathbf{1}\{b \in A, g \in B_n^\pm\}(\theta_g^{-1}\xi_A)(B)\lambda_L(dg)\lambda^*(db),$$

where we used L-invariance of A and $L^\perp$-invariance of B_n and B. Hence

$$\int_{A \cap B_n^\pm} (\theta_s^{-1}\xi_A)(B)\lambda(ds) = \lambda^*(A) \int \mathbf{1}\{g \in B_n^\pm\}(\theta_g^{-1}\xi_A)(B)\lambda_L(dg).$$

From (6.9) we conclude

$$\frac{\xi_A(\mathbf{1}_{A \cap B} * \mathbf{1}_{B_n^\pm})}{\lambda(A \cap B_n^\pm)} = \frac{1}{\delta(B_n^\pm)} \int_{B_n^\pm} (\theta_g^{-1}\xi_A)(B)\lambda_L(dg) = \frac{1}{\lambda_L(B_n^\pm \cap L)} \int_{B_n^\pm \cap L} (\theta_g^{-1}\xi_A)(B)\lambda_L(dg)$$

$$= \frac{1}{\lambda_L(B_n^\pm \cap L)} \int_{B_n^\pm \cap L} (\theta_g^{-1}\xi)(A \cap B)\lambda_L(dg)$$

$$= \frac{1}{\lambda_L(B_n^\pm \cap L)} \int_{B_n^\pm \cap L} f(\theta_g^{-1}\xi)\lambda_L(dg),$$

where we have put $f(\mu) := \mu(A \cap B)$. Here, we may apply the spatial ergodic Theorem 6.6 to the flow $\tilde{\theta}_z := \theta_z^{-1}, z \in \mathbb{Z}^d$, by replacing $\mathbb{R}^d$ by the k-dimensional space L and to the above function f, since either of the sequences $(B_n^\pm \cap L)$ consists of bounded, convex subsets of L with the property that $r_L(B_n^\pm \cap L) \to \infty$, $r_L(B_n^\pm \cap L)$ denoting the radius of the largest k-dimensional ball contained in $B_n^\pm \cap L$. This gives

$$\frac{\xi_A(\mathbf{1}_{A \cap B} * \mathbf{1}_{B_n^\pm})}{\lambda(A \cap B_n^\pm)} \to \mathbb{E}[\xi(A \cap B)|\mathcal{I}_\xi], \quad n \to \infty \quad a.s. \tag{6.10}$$

In addition, Theorem 6.6 also yields convergence in L^p whenever $f(\xi) = \xi(A \cap B) \in L^p$. The inequalities in (6.6) yield together with (6.7) and (6.10) that

$$\frac{\xi(A \cap B_n)}{\lambda(A \cap B_n)} \to \frac{\mathbb{E}[\xi(A \cap B)|\mathcal{I}_\xi]}{\lambda(A \cap B)} = \bar{\xi}_A, \quad n \to \infty,$$

in the respective sense under the corresponding condition. Since the approximating sequence is independent of B, this also holds for the limit. $\square$

6.3 Ergodicity and Cumulative Palm measure

The aim of this section is to show that the Cumulative Palm measure naturally arises in the limit of spatial averaging procedures under ergodicity assumptions. To accomplish that we shall use our two ergodic Theorems 6.3 and 6.13. We shall treat the case $\mathbb{Z}^d \hookrightarrow \mathbb{R}^d$ in Subsection 6.3.1 and the case $L \hookrightarrow \mathbb{R}^d$ in Subsection 6.3.2.

6.3.1 The grid-stationary case

For the action of $\mathbb{Z}^d$ on $\mathbb{R}^d$ we naturally choose the system of orbit representatives $O = [0, 1)^d$ such that $\beta(x)$ denotes the fractional part of $x \in \mathbb{R}^d$. We note that the inversion kernel κ (see Theorem 3.1) of this action is given by

$$\kappa_{\beta(x),x} = \delta_{x - \beta(x)}, \quad x \in \mathbb{R}^d.$$

We note that any measurable function $h : \Omega \times \mathbb{R}^d \to [0, \infty)$ satisfies condition (3.16) in this setting, since

$$\int h(\theta_g^{-1}\omega, \beta(x)) \kappa_{\beta(x),x}(dg) = h(\theta_{x-\beta(x)}^{-1}\omega, \beta(x)) < \infty, \quad x \in \mathbb{R}^d, \omega \in \Omega.$$

Thus, given a $\mathbb{Z}^d$-stationary random measure η in $\mathbb{R}^d$ we may form the h-transform ξ of η defined as in (3.17) via

$$\xi(\omega, \cdot) := \iint \mathbf{1}\{x \in \cdot\} h(\theta_g^{-1}\omega, \beta(x)) \kappa_{\beta(x),x}(dg)\eta(dx),$$

and note that ξ reduces to

$$\xi(\omega, \cdot) = \int \mathbf{1}\{x \in \cdot\} h(\theta_{x-\beta(x)}^{-1}\omega, \beta(x))\eta(dx).$$

Applying Theorem 6.2 to this h-transform yields the following result.

Corollary 6.14. (*h-transform convergence for $\mathbb{Z}^d$*) *Let η denote a $\mathbb{Z}^d$-stationary random measure in $\mathbb{R}^d$, A a $\mathbb{Z}^d$-invariant measurable subset of $\mathbb{R}^d$ and B_n a sequence of $\mathbb{Z}^d$-symmetric subsets of $\mathbb{R}^d$ such that $B_n \cap \mathbb{Z}^d$ is an increasing sequence of boxes. Then, for any measurable $h : \Omega \times \mathbb{R}^d \to [0, \infty)$ satisfying*

$$\int_{[0,1)^d \cap A} h(\theta_{x-\beta(x)}^{-1}, \beta(x))\eta(dx) \in L \log^{d-1} L(\mathbb{P})$$

its holds that a.s.

$$\frac{1}{\lambda(A \cap B_n)} \int_{A \cap B_n} h(\theta_{x-\beta(x)}^{-1}, \beta(x))\eta(dx) \to \frac{1}{\lambda(A \cap [0, 1)^d)} \mathbb{E}\left[\int_{A \cap [0,1)^d} h(\theta_{x-\beta(x)}^{-1}, \beta(x))\eta(dx) \Big| \tilde{\mathcal{I}}\right]$$

for a σ-algebra $\tilde{\mathcal{I}} \subset \mathcal{I}_\eta = \{\eta^{-1}A : A \in \mathcal{M}(\mathbb{R}^d)\}$.

Proof. We may apply Theorem 6.3 to the $\mathbb{Z}^d$-stationary (see Lemma 3.18) random measure

$$\xi(\cdot) = \int \mathbf{1}\{x \in \cdot\} h(\theta_{x-\beta(x)}^{-1}, \beta(x))\eta(dx).$$

Then it remains to note that $\tilde{\mathcal{I}} := \mathcal{I}_\xi$ is contained in $\mathcal{I}_\eta$ by Lemma 3.18. $\qquad \square$

This leads to the following ergodic theorem exhibiting integrals with respect to the Cumulative Palm measure of a $\mathbb{Z}^d$-stationary and $\mathbb{Z}^d$-ergodic random measure η in $\mathbb{R}^d$ as a.s. limits of spatial integrals with respect to η over $\mathbb{Z}^d$-symmetrically increasing domains.

Theorem 6.15 (cumulative Palm measure and $\mathbb{Z}^d$-ergodicity). *Let η denote a $\mathbb{Z}^d$-stationary and $\mathbb{Z}^d$-ergodic random measure in $\mathbb{R}^d$. Then for a set A and a sequence (B_n) as in Theorem 6.3 and any measurable $h : \Omega \times S \to [0, \infty)$ as in Corollary 6.14 it holds that*

$$\frac{1}{\delta(B_n)} \int_{A \cap B_n} h(\theta^{-1}_{x-\beta(x)}, \beta(x)) \eta(dx) \to \int h(\omega, b) \mathbf{1}_A(b) \mathbb{Q}^\eta(d(\omega, b)) \quad a.s.$$

where β and the cumulative Palm measure $\mathbb{Q}^\eta$ are both with respect to $O = [0, 1)^d$.

Proof. By the above Corollary 6.14 and (2.13) it holds a.s.

$$\frac{1}{\lambda^*(A)\delta(B_n)} \int_{A \cap B_n} h(\theta^{-1}_{x-\beta(x)}, \beta(x)) \eta(dx) \longrightarrow \frac{\mathbb{E}\left[\int_{A \cap [0,1)^d} h(\theta^{-1}_{x-\beta(x)}, \beta(x)) \eta(dx) | \tilde{\mathcal{I}}\right]}{\lambda^d(A \cap [0, 1)^d)}$$

and since η is ergodic, this random limit equals a.s.

$$\frac{1}{\lambda^d(A \cap [0, 1)^d)} \mathbb{E}\left[\int_{A \cap B} h(\theta^{-1}_{x-\beta(x)}, \beta(x)) \eta(dx)\right].$$

By Lemma 4.13 and the calculation in Example 4.14 we may write this as

$$\frac{1}{\lambda^d(A \cap [0, 1)^d)} \int h(\omega, b) \mathbf{1}\{b \in A\} \mathbb{Q}(d(\omega, b)),$$

which yields the assertion after multiplying with $\lambda^*(A) = \lambda^d(A \cap [0, 1)^d)$. $\square$

6.3.2 The subspace-stationary case

To emphasize the analogy with the results in the previous section, we decided to formulate all results and proofs in this subsection in a copy-paste manner. We consider now the action of a k-dimensional subspace L of $\mathbb{R}^d$ ($0 \le k \le d$) on $\mathbb{R}^d$ via translation, and write in short $L \hookrightarrow \mathbb{R}^d$ for this operation. Note that the inversion kernel κ of this operation with respect to any chosen system of orbital representatives is nothing but

$$\kappa_{\beta(x),x} = \delta_{x-\beta(x)}, \quad x \in \mathbb{R}^d.$$

Any measurable function $h : \Omega \times \mathbb{R}^d \to [0, \infty)$ satisfies condition (3.16) in this setting as

$$\int h(\theta^{-1}_g \omega, \beta(x)) \kappa_{\beta(x),x}(dg) = h(\theta^{-1}_{x-\beta(x)} \omega, \beta(x)) < \infty, \quad x \in \mathbb{R}^d, \omega \in \Omega.$$

Thus, given an L-stationary random measure η in $\mathbb{R}^d$, we may h-transform η into the L-stationary random measure ξ defined as in (3.17) via

$$\xi(C) := \iint \mathbf{1}\{x \in C\} h(\theta^{-1}_g, \beta(x)) \kappa_{\beta(x),x}(dg) \eta(dx),$$

which reduces to

$$\xi(C) = \int \mathbf{1}\{x \in C\} h(\theta^{-1}_{x-\beta(x)}, \beta(x)) \eta(dx).$$

Using this transformation we derive:

Corollary 6.16. (*h*-transform convergence for linear subspaces) *Consider an L-stationary random measure η in $\mathbb{R}^d$. Then if $A \subset \mathbb{R}^d$ is L-invariant and admissible and $B_1 \subset B_2 \subset \ldots$ is a nested sequence of convex L-symmetric and $L^\perp$-invariant sets in $\mathbb{R}^d$ with $\delta(B_n) \to \infty$, then for any measurable $h : \Omega \times S \to [0, \infty)$ it holds a.s.*

$$\frac{1}{\lambda(A \cap B_n)} \int_{A \cap B_n} h(\theta_{p_L(x)}^{-1}, p_{L^\perp}(x))\eta(dx) \to \frac{1}{\lambda(A \cap B)} \mathbb{E}\left[\int_{A \cap B} h(\theta_{p_L(x)}^{-1}, p_{L^\perp}(x))\eta(dx) \Big| \tilde{\mathcal{I}} \right]$$

for a σ-algebra $\tilde{\mathcal{I}} \subset \mathcal{I}_\eta = \{\eta^{-1}A : A \in \mathcal{M}(\mathbb{R}^d)\}$, where $p_L(x)$ and $p_{L^\perp}(x)$ denote the orthogonal projections of x on L and $L^\perp$ respectively.

Proof. We may apply Theorem 6.13 to the L-stationary (see Lemma 3.18) random measure

$$\xi(\cdot) = \int \mathbf{1}\{x \in \cdot\} h(\theta_{x-\beta(x)}^{-1}, \beta(x))\eta(dx).$$

Then it remains to note that $\tilde{\mathcal{I}} := \mathcal{I}_\xi$ is contained in $\mathcal{I}_\eta$ by Lemma 3.18 and that $x - \beta(x) = p_L(x), x \in \mathbb{R}^d$, and $\beta(x) = p_{L^\perp}(x), x \in \mathbb{R}^d$. $\qquad\square$

This leads to the following ergodic theorem exhibiting integrals with respect to the Cumulative Palm measure of an L-stationary ergodic random measure η in $\mathbb{R}^d$ as a.s. limits of spatial integrals with respect to η over L-symmetrically increasing domains:

Theorem 6.17 (cumulative Palm measure and subspace-ergodicity). *Let η denote an L-stationary random measure on $\mathbb{R}^d$ which is ergodic. Then for a set A and a sequence (B_n) as in Theorem 6.13 and any measurable $h : \Omega \times S \to [0, \infty)$*

$$\frac{1}{\delta(B_n)} \int_{A \cap B_n} h(\theta_{p_L(x)}^{-1}, p_{L^\perp}(x))\eta(dx) \to \int h(\omega, b)\mathbf{1}\{b \in A\}\mathbb{Q}(d(\omega, b)) \quad a.s.,$$

where $p_L(x)$ and $p_{L^\perp}(x)$ denote the orthogonal projections of x on L and $L^\perp$ respectively.

Proof. By the above Corollary 6.16 and (2.13) we have

$$\frac{1}{\lambda^*(A)\delta(B_n)} \int_{A \cap B_n} h(\theta_{x-\beta(x)}^{-1}, \beta(x))\eta(dx) \longrightarrow \frac{1}{\lambda^*(A)\delta(B)} \mathbb{E}\left[\int_{A \cap B} h(\theta_{x-\beta(x)}^{-1}, \beta(x))\eta(dx) | \tilde{\mathcal{I}} \right]$$

a.s. and since η is ergodic this random limit equals a.s.

$$\frac{1}{\lambda^*(A)\delta(B)} \mathbb{E}\left[\int_{A \cap B} h(\theta_{x-\beta(x)}^{-1}, \beta(x))\eta(dx) \right].$$

By Lemma 4.13 and the calculation in Example 4.14 we may write this as

$$\frac{1}{\lambda^*(A)} \int h(\omega, b)\mathbf{1}\{b \in A\}\mathbb{Q}(d(\omega, b)),$$

which yields the assertion after multiplying with $\lambda^*(A)$. $\qquad\square$

Chapter 7

On some new models in Stochastic Geometry

In this chapter we give several applications of the results obtained in Chapters 4, 5 and 6. It should be mentioned at this point that the use of Palm methods in Stochastic Geometry began with the seminal paper [47] by Joseph Mecke, while Meijering [49] seems to be the first who investigated a random geometric model under ergodicity assumptions. He was then followed by others such as Ambartzumian [3, 4], Miles [50, 51] and Cowan [12, 13]. The mass-transport principle in the form of Theorem 5.6 has certainly been implicitly used in the transitive unimodular special case whenever Neveu's exchange formula was used, but even in this special case the intuition of transporting mass seems to be new until recently [39]. Also it seems like it has never been used in its integrated form derived here in Theorems 5.2 and 5.5.

Section 7.1 is on *random tessellations*, where the central result is Theorem 7.8. It gives a structurally quite explicit expression for the quasi-distribution of the typical cell of a Cox-Delaunay mosaic, seen from the center of the unique ball in which all its vertices are contained. Then, we use the Palm MTP (Theorem 5.6) to identify suitably defined 0-cells of random partitions on Riemannian manifolds as volume-weighted versions of suitably defined typical cells in Section 7.2. The use of the integrated version of the MTP (Theorem 5.5) is then illustrated in Section 7.3 where we give two applications. One is on approximation of Borel sets with random partitions and the other on the intensity measure of the restriction of k-dimensional Hausdorff measure to the k-skeleton of a random tessellation. Finally, in the last Section 7.4 of this thesis, we quickly illustrate the use of our results on group ergodic random measures in Chapter 6 by giving applications.

7.1 Random tessellations

After introducing the relevant object of this section in Subsection 7.1.1, we proceed in Subsection 7.1.2 with an investigation of several cumulative Palm measures derived from an arbitrary random tessellation. In Subsection 7.1.3 we consider a simple Cox process ξ in $\mathbb{R}^d$ that is stationary with respect to a subgroup G of the group of rigid motions G_d. This includes e.g. the cases $G = SO(d)$ or $G = L$ where L is a k-dimensional linear subspace of $\mathbb{R}^d$ where $0 \leq k \leq d$. Note that here $k = d$ is the completely stationary case while $u = 0$ is the completely non-stationary case. This Cox process induces a (random) Delaunay tessellation and we shall give an explicit formula for the distribution of the typical cell of such a Cox-Delaunay tessellation under all cells lying in a fixed invariant class. Examples of these invariant classes include the following three: Fixing a G-invariant set $A \subset \mathbb{R}^d$ the set of all cells contained in A does not change under shifts induces by G, just as the set of all cells having a center in A, or all cells hitting A. These results are the content of Subsection 7.1.4.

7.1.1   Tessellations

A *tessellation* or *mosaic* in $\mathbb{R}^d$ (cf. [64, Section 10.1]) is a countable system $\mathbf{m}$ of subsets satisfying the conditions

(i) $\mathbf{m}$ is a with respect to the Fell topology (see [64, Definition 2.1.1]) locally finite system of non-empty closed sets.

(ii) The sets $K \in \mathbf{m}$ are compact, convex and have interior points.

(iii) $\bigcup_{K \in \mathbf{m}} K = \mathbb{R}^d$.

(iv) If $K, K' \in \mathbf{m}$ with $K \neq K'$, then $\operatorname{int} K \cap \operatorname{int} K' = \emptyset$.

The elements of such a mosaic are also called *cells* (of $\mathbf{m}$) and they are convex polytopes ([64, Lemma 10.1.1]). A *face* of a convex polytope P is the intersection of P with any of its supporting hyperplanes, and if this intersection is of dimension k, the face is called k-*face*. The cells themselves are consistently also called d-faces. 0-faces are also called *vertices* (identifying $\{x\}$ with x), 1-faces are the *edges* while $d - 1$-faces are called *facets* of P. Given a mosaic $\mathbf{m}$ and a polytope $P \in \mathbf{m}$, we denote by $\mathcal{F}_k(P)$ the set of all k-faces of P and by $\mathcal{F}_k(\mathbf{m})$ we denote $\bigcup_{P \in \mathbf{m}} \mathcal{F}_k(P)$. It is convenient to write $\mathcal{F}(P) = \bigcup_{0 \leq k \leq d} \mathcal{F}_k(P)$ and similarly $\mathcal{F}(\mathbf{m}) = \bigcup_{0 \leq k \leq d} \mathcal{F}_k(\mathbf{m})$. We call a mosaic $\mathbf{m}$ *face-to-face* if

$$P \cap P' \in (\mathcal{F}(P) \cap \mathcal{F}(P')) \cup \{\emptyset\}, \quad P, P' \in \mathbf{m}.$$

Both the set $\mathbb{M}$ of all mosaics and the set $\mathbb{M}^*$ of all face-to-face mosaics are Borel subsets in the space of all closed subsets of the space of all closed non-empty subsets of $\mathbb{R}^d$ (each time considering the Fell topology), see [64, Lemma 10.1.2]. A *particle process* in $\mathbb{R}^d$ is a point process in the space of all non-empty compact subsets of $\mathbb{R}^d$, where this space is endowed with the trace topology resp. σ-field from the surrounding space of all closed subsets of $\mathbb{R}^d$. A *random mosaic* in $\mathbb{R}^d$ is now a particle process X in $\mathbb{R}^d$ satisfying $\mathbb{P}(X \in \mathbb{M}^*) = 1$. Thus, random mosaics are per definition a.s. face-to-face.

Given a compact subset C in $\mathbb{R}^d$, we may assign a center to it in a G_d-covariant manner in several ways, where G_d denotes the group of rigid motions in $\mathbb{R}^d$. E.g. we may assign to C the center of the uniquely determined circumball of C (the smallest ball containing C). We may even allow for additional randomness and consider *generalized center functions* $\pi : \Omega \times \mathcal{C}' \to \mathbb{R}^d$ satisfying

$$\pi(\theta_\varphi \omega, \varphi(C)) = \varphi(\pi(\omega, C)), \quad C \in \mathcal{C}', \varphi \in G_d, \omega \in \Omega.$$

Now, given a random mosaic X in $\mathbb{R}^d$, we may consider for any $0 \leq k \leq d$ the point process

$$N_k(\omega, \cdot) := \sum_{F \in \mathcal{F}_k(X(\omega))} \delta_{\pi(\omega, F)}(\cdot), \quad \omega \in \Omega, \tag{7.1}$$

of the centers of the k-faces and assume that π is such that N_k is a.s. simple for each $0 \leq k \leq d$. In addition, we define the random measure

$$M_k(\omega, \cdot) := \sum_{F \in \mathcal{F}_k(X(\omega))} \mathcal{H}^k(F \cap \cdot), \quad \omega \in \Omega, \tag{7.2}$$

where $\mathcal{H}^k$ denotes the k-dimensional Hausdorff measure in $\mathbb{R}^d$. Further, if t is a point in the relative interior of a k-face $F \in \mathcal{F}_k(X(\omega))$ in configuration $\omega \in \Omega$, we write $\pi_k(\omega, t) := \pi(\omega, F)$.

7.1.2   Derived cumulative Palm measures

Given a G-stationary random tessellation X on $\mathbb{R}^d$, where G is some closed unimodular subgroup of the group of rigid motions G_d of $\mathbb{R}^d$, we will compare the cumulative Palm measures of M_k and N_k with respect to some fixed measurable system O of orbit representatives in this subsection. The relation is well-known for completely stationary random mosaics, see e.g. [5].

Lemma 7.1. (cumulative Palm measures of M_k and N_k) *Let X denote a G-stationary random tessellation of $\mathbb{R}^d$, where G denotes some closed unimodular subgroup of G_d. Let further N_k and M_k be defined as in (7.1) and (7.2) respectively. Then the cumulative Palm measures $\mathbb{Q}^{N_k}$ and $\mathbb{Q}^{M_k}$ with respect to an arbitrary measurable system O of orbit representatives satisfy*

$$\int \mathbf{1}\{(\omega, b) \in \cdot\}\mathcal{H}^k(C_k(\omega, b))\mathbb{Q}^{N_k}(d(\omega, b)) \tag{7.3}$$

$$= \iint \mathbf{1}\{(\theta_g^{-1}\omega, \beta(\pi_k(\omega, b))) \in \cdot\}\kappa_{\beta(\pi_k(\omega,b)),\pi_k(\omega,b)}(dg)\mathbb{Q}^{M_k}(d(\omega, b)),$$

and

$$\iint \mathbf{1}\{(\theta_g^{-1}\omega, \beta(\pi_k(\omega, b))) \in \cdot\}\frac{1}{\mathcal{H}^k(V_k(\omega, b))}\kappa_{\beta(\pi_k(\omega,b)),\pi_k(\omega,b)}(dg)\mathbb{Q}^{M_k}(d(\omega, b))$$

$$= \int \mathbf{1}\{(\omega, b) \in \cdot\}\mathbb{Q}^{N_k}(d(\omega, b)). \tag{7.4}$$

Proof. We notice that

$$\iint \mathbf{1}\{(s, t) \in \cdot\}\mathbf{1}\{t \in C^k(\omega, s)\}\mathcal{H}^k(dt)N_k(ds)$$

$$= \iint \mathbf{1}\{(\pi_k(t), t) \in \cdot\}\mathbf{1}\{t \in C^k(\omega, s)\}\mathcal{H}^k(dt)N_k(ds)$$

$$= \int \mathbf{1}\{(\pi_k(t), t) \in \cdot\}M_k(dt)$$

$$= \iint \mathbf{1}\{(s, t) \in \cdot\}\delta_{\pi_k(\omega,t)}(ds)M_k(dt).$$

Thus we may put in the Palm MTP (Theorem 5.6) $\xi = N_k$, $\eta = M_k$, $\gamma(\omega, s, \cdot) = \mathcal{H}^k(C_k(\omega, s) \cap \cdot)$ and $\delta(\omega, t, \cdot) = \delta_{\pi_k(\omega, t)}(\cdot)$, and receive

$$\iint m(\omega, b, t)\mathbf{1}\{t \in C_k(\omega, b)\}\mathcal{H}^k(dt)\mathbb{Q}^{N_k}(d(\omega, b)) = \int m(\omega, \pi_k(\omega, b), b)\mathbb{Q}^{M_k}(d(\omega, b)),$$

for any jointly G-invariant measurable $m : \Omega \times \mathbb{R}^d \times \mathbb{R}^d \to [0, \infty)$. Choosing here the jointly G-invariant

$$m(\omega, s, t) = \int \mathbf{1}\{(\theta_g^{-1}\omega, g^{-1}s) \in \cdot\}\kappa_{\beta(s),s}(dg), \quad \omega \in \Omega, s, t \in \mathbb{R}^d,$$

yields

$$\iiint \mathbf{1}\{(\theta_g^{-1}\omega, b) \in \cdot\}\mathcal{H}^k(C_k(\omega, b))\kappa_{b,b}(dg)\mathbb{Q}^{N_k}(d(\omega, b))$$

$$= \iint \mathbf{1}\{(\theta_g^{-1}\omega, \beta(\pi_k(\omega, b))) \in \cdot\}\kappa_{\beta(\pi_k(\omega,b)),\pi_k(\omega,b)}(dg)\mathbb{Q}^{M_k}(d(\omega, b)).$$

Since

$$\mathcal{H}^k(C_k(\omega, b)) = \mathcal{H}^k(g^{-1}C_k(\omega, b)) = \mathcal{H}^k(C_k(\theta_g^{-1}\omega, b)), \quad g \in G_{b,b}, b \in O, \omega \in \Omega,$$

the left-hand side reduces by (4.3) to the left-hand side of (7.3)

$$\int \mathbf{1}\{(\omega, b) \in \cdot\}\mathcal{H}^k(C_k(\omega, b))\mathbb{Q}^{N_k}(d(\omega, b)).$$

Equation (7.4) follows from the same arguments when using instead

$$m(\omega, s, t) = \mathbf{1}\{\mathcal{H}^k(C_k(\omega, s)) > 0\}\frac{1}{\mathcal{H}^k(C_k(\omega, s))}\int \mathbf{1}\{(\theta_g^{-1}\omega, g^{-1}s) \in \cdot\}\kappa_{\beta(s),s}(dg),$$

where $\omega \in \Omega, s, t \in \mathbb{R}^d$ (with the usual convention $0 \cdot \infty = 0$). □

Until the end of this subsection, X denotes a $\mathbb{Z}^d$-stationary tessellation. We may define for any fixed $\mathbb{Z}^d$-invariant set A with $(\mathbb{E}N_i)^*(A) < \infty$ for each $i \in \{0, 1, \ldots, d\}$ the *i-cell density* on A as

$$\gamma^{(i)}(A) := (\mathbb{E}N_i)^*(A), \quad i \in \{0, 1, \ldots, d\}.$$

We note that, introducing a G-symmetric set B of width $\delta(B) = 1$, we may write by (2.12)

$$\gamma^{(i)}(A) = \mathbb{E}N_i(A \cap B), \quad i \in \{0, 1, \ldots, d\}.$$

We call

$$\gamma^{(i)} := \gamma^{(i)}(\mathbb{R}^d) \tag{7.5}$$

simply the *i-cell density* of X, and define similarly as in Definition 4.23 the probability measures

$$\mathbb{P}^{N_i}(\cdot) := \frac{1}{\gamma^{(i)}}\mathbb{Q}^{N_i}(\cdot), \quad i \in \{0, \ldots, d\}.$$

We further define

$$n_{ij} := \iint \mathbf{1}\{C_i(\omega, b) \subset C_j(\omega, t)\} N_j(\omega, dt) \mathbb{P}^{N_i}(d(\omega, b)), \quad i \leq j, \qquad (7.6)$$

and

$$n_{ij} := \iint \mathbf{1}\{C_j(\omega, t) \subset C_i(\omega, b)\} N_j(\omega, dt) \mathbb{P}^{N_i}(d(\omega, b)), \quad i > j, \qquad (7.7)$$

and interpret n_{ij} as the mean number of j-faces containing the typical i-face for $i \leq j$, while for $i > j$ the quantity n_{ij} may be interpreted as the mean number of j-faces contained in the typical i-face.

The following result about $\mathbb{Z}^d$-stationary tessellations X stems from a similar balancing procedure, using N_i and N_j for (different) $i, j \in \{0, 1, \ldots, d\}$ instead of balancing N_k and M_k. It extends aspects of [5, Proposition 2.2] and [42, Theorem 1] to a partially stationary setting.

Lemma 7.2 (mean relations for numbers of i-faces). *Let X denote a $\mathbb{Z}^d$-stationary random tessellation of $\mathbb{R}^d$. Further assume that all $\gamma^{(i)}$ defined in (7.5) are finite, just as the n_{ij} defined as in (7.6) and (7.7) for fixed $i, j \in \{0, 1, \ldots, d\}$. Then*

$$\gamma^{(i)} n_{ij} = \gamma^{(j)} n_{ji}. \qquad (7.8)$$

with respect to some fixed measurable system O of orbit representatives.

Proof. By symmetry we may assume without loss of generality that $i \leq j$. The assertion clearly reduces to

$$\iint \mathbf{1}\{C_i(\omega, b) \subset C_j(\omega, t)\} N_j(\omega, dt) \mathbb{Q}^{N_i}(d(\omega, b))$$
$$= \iint \mathbf{1}\{C_i(\omega, s) \subset C_j(\omega, b)\} N_i(\omega, ds) \mathbb{Q}^{N_j}(d(\omega, b)),$$

which is just the Palm MTP 5.6 when using $m(\omega, s, t) = \mathbf{1}\{C_i(\omega, s) \subset C_j(\omega, t)\}$, $\xi = N_i$, $\gamma = N_j$, $\eta = N_j$ and $\delta = M_i$. $\qquad \square$

In the above proof, intuitively each i-face sends mass 1 to its adjacent j-faces. The idea in the following result is that each i-face transports its internal angle to the d-faces in which it is contained. Here, given a polytope P and a face F of this polytope, the *internal angle* of P at F is defined by

$$\beta(F, P) := \frac{\lambda^d(\text{Cone}(P, F) \cap B^d)}{\kappa_d},$$

where $\text{Cone}(P, F)$ is the cone spanned by P at an arbitrary relatively interior point z of F. More precisely,

$$\text{Cone}(P, F) = \{\alpha(x - z) : x \in P, \alpha \geq 0\}.$$

The following result extends [64, Theorem 10.1.3].

Lemma 7.3 (transporting internal angles). *Let X denote a $\mathbb{Z}^d$-stationary random mosaic on $\mathbb{R}^d$ and let $g : \mathcal{C}' \to [0, \infty)$ be $\mathbb{Z}^d$-invariant and measurable. Then for any $j \in \{0, \ldots, d\}$*

$$\gamma^{(j)} \int g(C_j(\omega, b))\mathbb{P}^{N_j}(d(\omega, b)) = \gamma^{(d)} \int \sum_{F \in \mathcal{F}_j(C_d(\omega,b))} \beta(F, P)g(F)\mathbb{P}^{N_d}(d(\omega, b)).$$

In particular

$$\gamma^{(j)} = \gamma^{(d)} \int \sum_{F \in \mathcal{F}_j(C_d(\omega,b))} \beta(F, P)\mathbb{P}^{N_d}(d(\omega, b)),$$

and in addition

$$\sum_{i=0}^{d}(-1)^i\gamma^{(i)} = 0. \tag{7.9}$$

Proof. The first equation again follows from the Palm MTP Theorem 5.6, using $\xi = N_i$, $\gamma = N_d$, $\eta = N_d$, $\delta = N_i$ and

$$m(\omega, s, t) = \mathbf{1}\{C_i(\omega, s) \subset C_d(\omega, t)\}\beta(C_i(\omega, s), C_d(\omega, t))g(C_i(\omega, s)).$$

The second equation is the special case $g \equiv 1$, and the last follows from alternatingly summing up the second equation for $j = 0, \ldots, d$ and then using Fubini and *Gram's relation* (see [23])

$$\sum_{i=0}^{d}(-1)^i \sum_{F \in \mathcal{F}_i(P)} \beta(F, P) = 0,$$

which holds for any d-dimensional polytope. $\square$

Remark 7.4. Equation (7.9) is an *Euler type relation* for $\mathbb{Z}^d$-stationary random mosaics in $\mathbb{R}^d$. In $d = 2$, we obtain together with (7.8) for $(i, j) = (0, 2)$ and $(i, j) = (1, 2)$, the system of linear equations (using $n_{12} = 2$ and $n_{21} = n_{20}$)

$$\gamma^{(0)}n_{02} = \gamma^{(2)}n_{20}$$
$$\gamma^{(1)}2 = \gamma^{(2)}n_{20}$$
$$\gamma^{(0)} - \gamma^{(1)} + \gamma^{(2)} = 0.$$

It readily implies

$$n_{20} = \frac{2n_{02}}{n_{02} - 2}.$$

For a $\mathbb{Z}^2$-stationary normal tessellation in $\mathbb{R}^2$ (where *normal* means that each vertex is contained in 3 cells and each edge is contained in 2 cells) we have $n_{02} = 3$ and the result then implies

$$n_{20} = 6.$$

A concrete example is a $\mathbb{Z}^d$-stationary Poisson process whose intensity measure is absolutely continuous with respect to Lebesgue measure.

7.1.3 Voronoi and Delaunay tessellations

Any locally finite set $A \subset \mathbb{R}^d$ induces a *Voronoi tessellation (or Voronoi mosaic)* of $\mathbb{R}^d$ defined as the collection of *Voronoi cells*

$$C(A, s) := \{x \in \mathbb{R}^d : ||x - s|| \leq ||x - y||, y \in A\}, \quad s \in A,$$

and we put $C(A, s) := \emptyset$ if $s \notin A$. It is known that if $\operatorname{conv} A = \mathbb{R}^d$ ('conv' denoting the convex hull operator) then all Voronoi cells are bounded (the converse fails in general), see [64, p. 471]. Further, it is known that if all Voronoi cells induced by a locally finite, non-empty set A are bounded, then the Voronoi tessellation is a face-to-face mosaic (see Subsection 7.3.2 for definitions and [64, Theorem 10.2.1] for a proof of this assertion). In addition, if the points of A are in *general position*, i.e. no $(d+1)$ of them lie in a $d-1$-dimensional affine subspace of $\mathbb{R}^d$, and if any $d+2$ of them are not located on a sphere, then the Voronoi mosaic is normal. Here, a face-to-face mosaic is called *normal* if each of its k-faces is contained in the boundary of precisely $d - k + 1$ cells. If the generating set A is random, e.g. given by the support of a simple point process ξ, the induced tessellation is random, too. Random (or non-random) Voronoi tessellations are of considerable interest both in theory and application. It is clear that they may be constructed in more general metric spaces. Comprehensive and detailed overviews on this subject are given in [58, 66, 64, 54]. Earlier contributions in this field are [52, 53, 48] while [55] and [27] ([27] considers instead of $\mathbb{R}^d$ the 3-dimensional hyperbolic space) are more recent papers.

If $\mathbf{m}$ denotes the Voronoi mosaic generated by some locally finite subset A of $\mathbb{R}^d$, then we define for $s \in \mathcal{F}_0(\mathbf{m})$

$$D(s, A) := \operatorname{conv}\{x \in A : s \in \mathcal{F}_0(C(A, x))\}.$$

The collection of all these sets, where s ranges over the vertices of $\mathbf{m}$, is called the *Delaunay mosaic* generated by A. Again, this mosaic is face-to-face if $\operatorname{conv} A = \mathbb{R}^d$ (see [64, Theorem 10.2.6]). Also, if the points of A are in *general position* and if any $d + 2$ of them are not located on a sphere, then it is *simplicial* in the sense that every cell of it has $d + 1$ vertices (and is thus a d-simplex).

We want to consider *random* Delaunay mosaics, where the generating set A is the support of a simple Cox process ξ. Such random mosaics will be called *Cox-Delaunay mosaics*. We need to put a few regularity conditions on the Cox process ξ, namely we shall assume that

 (i) ξ is a.s. simple (which is the case if and only if η is a.s. diffuse, as is easy to see),

 (ii) $\operatorname{conv} \xi = \mathbb{R}^d$ a.s. (where we identify as usual ξ with its support),

 (iii) the points of ξ are a.s. in general position,

 (iv) no $d + 2$ of the points of ξ lie on a sphere.

We call a Cox process ξ on $\mathbb{R}^d$ *regular*, if it satisfies (i)-(iv). These assumptions are not too restrictive and allow many interesting cases. For instance, any Cox process ξ driven by a random measure η of the form

$$\eta(\omega, \cdot) = \int \mathbf{1}\{s \in \cdot\} f(\omega, s) \lambda^d(ds), \quad \omega \in \Omega,$$

where $f : \Omega \times \mathbb{R}^d \to [c, \infty)$ (for some $c > 0$) is measurable and such that $f(\omega, \cdot)$ is continuous for $\mathbb{P}$-a.e. ω, is regular. Another regular example is a Cox process driven by the random measure M_k (for any $0 < k \leq d$) based on any random tessellation X of $\mathbb{R}^d$.

From what has been said, it follows that the Delaunay mosaic X based on a regular Cox process ξ is a random (face-to-face) mosaic, which is a.s. simplicial.

7.1.4   Typical cells of Cox-Delaunay tessellations

Let L denote a fixed k-dimensional linear subspace of $\mathbb{R}^d$ where $1 \leq k \leq d$ and let ξ be a regular Cox process in $\mathbb{R}^d$ driven by an L-stationary random measure η (note that the completely stationary case $k = d$ is included). X will denote the random Cox-Delaunay mosaic induced by ξ. Let $\Delta^{(d)}$ denote the space of all d-dimensional simplices in $\mathbb{R}^d$, endowed with the trace topology and σ-algebra inherited from the space of all closed subsets of $\mathbb{R}^d$. We may choose an arbitrary L-covariant deterministic (!) measurable center function $z : \Delta^{(d)} \to \mathbb{R}^d$, which we assume to be a function of the $d+1$ vertices, such that we may write $z(K) = z(x_0, \ldots, x_d) = z(x)$, putting $x = (x_0, \ldots, x_d)$. Examples are the center of the unique circumball of a cell and the center of the unique ball having all the $d + 1$ vertices of the simplex in its boundary. Following [64, p. 495] we write for $K \in \Delta^{(d)}$ its vertices $x_0 = x_0(K), \ldots, x_d = x_d(K)$ in lexicographic order and as such as measurable functions of K. Given such vertices $x_0, \ldots, x_d$ we write $B^d(x_0, \ldots, x_d)$ for the unique open ball having these vertices in its boundary. It is not difficult to show that the point process of centers

$$\zeta := \sum_{K \in X} \delta_{z(K)} \tag{7.10}$$

is a.s. simple. Given in configuration ω a center $s \in \zeta(\omega)$, it is for a.e. configuration ω the center of a unique cell for which we write $C(\omega, s)$. If $s \notin \zeta(\omega)$, we put $C(\omega, s) = \emptyset$.

As the Cox-Delaunay mosaic X is a deterministic function of ξ, whose distribution in turn is fully determined by η, the distributions of all objects derived from ξ must also depend on η alone. In particular, this must be true for the expression

$$\mathbb{Q}_\zeta^I(C(\theta_e, \mathbf{b}) \in \cdot),$$

where $\mathbf{b} : O \to O$ is the identity on $O := L^\perp$.

Lemma 7.5 (Cox-Delaunay cells). *Let L denote a k-dimensional linear subspace of $\mathbb{R}^d$ ($1 \leq k \leq d$) and consider its canonical action $L \hookrightarrow \mathbb{R}^d$ via translation. Given a Cox-Delaunay mosaic in $\mathbb{R}^d$ induced by a regular Cox process driven by an L-stationary random measure η, we have with the above notations*

$$\mathbb{Q}^\zeta((C(\theta_e, \mathbf{b}), \mathbf{b}) \in \cdot) = \frac{1}{(d+1)!} \mathbb{E} \int \mathbf{1}\{(\mathrm{conv}\{x_0, \ldots, x_d\} - z(x) + \beta(z(x)), \beta(z(x))) \in \cdot\}$$

$$\times \mathbf{1}_B(z(x)) e^{-\eta(B^d(x_0, \ldots, x_d))} \eta^{d+1}(dx),$$

where $\mathbb{Q}^\zeta$ is the cumulative Palm measure of the center process ζ with respect to $O = L^\perp$ and B is any L-symmetric set with $\delta(B) = 1$.

Proof. Choosing $w = \mathbf{1}_B$ in (4.4) yields

$$\mathbb{Q}^\zeta((C(\theta_e, \mathbf{b}), \mathbf{b}) \in \cdot) = \mathbb{E} \iint \mathbf{1}\{(g^{-1}C(\theta_e, s), \beta(s)) \in \cdot\} \kappa_{\beta(s), s}(dg) \mathbf{1}_B(s)\zeta(ds),$$

where we also used L-covariance of C. Let ξ denote the regular Cox process driven by η. As $\kappa_{\beta(s), s} = \delta_{s - \beta(s)}, s \in \mathbb{R}^d$, the right-hand side equals

$$\frac{1}{(d+1)!} \mathbb{E} \sum_{(x_0, \ldots, x_d) \in \xi^{(d+1)}} \mathbf{1}\{(\operatorname{conv}\{x_0, \ldots, x_d\} - z(x) + \beta(z(x)), \beta(z(x)) \in \cdot\}$$
$$\times \mathbf{1}_B(z(x)) \mathbf{1}\{\xi(B^d(x_0, \ldots, x_d)) = 0\},$$

and using the multivariate Cox formula (4.15), this may be written as

$$\frac{1}{(d+1)!} \mathbb{E} \int \mathbf{1}\{(\operatorname{conv}\{x_0, \ldots, x_d\} - z(x) + \beta(z(x)), \beta(z(x))) \in \cdot\} \mathbf{1}_B(z(x))$$
$$\times \mathbf{1}\{\xi(B^d(x_0, \ldots, x_d)) = 0\} \eta^{d+1}(dx),$$

since

$$\left(\xi + \sum_{i=0}^d \delta_{x_i}\right)(B^d(x_0, \ldots, x_d)) = \xi\left(B^d(x_0, \ldots, x_d)\right)$$

($B^d(x_0, \ldots, x_d)$ is open and thus disjoint with its boundary). Here we may use Lemma 2.23 to write this as

$$\frac{1}{(d+1)!} \mathbb{E} \int \mathbf{1}\{(\operatorname{conv}\{x_0, \ldots, x_d\} - z(x) + \beta(z(x)), \beta(z(x))) \in \cdot\} \mathbf{1}_B(z(x))$$
$$\times \int \mathbf{1}\{\mu(B^d(x_0, \ldots, x_d)) = 0\} \mathbb{P}(\xi \in d\mu|\eta) \eta^{d+1}(dx),$$

which is the assertion since $\mathbb{P}(\xi \in \cdot|\eta)$ is a.s. the law of a Poisson process with intensity measure η. $\qquad\square$

Working in $\mathbb{R}^d$ has the advantage that we may use the group structure of $O = L^\perp$ as well, to further simplify the above result. Namely, we may look at a *centralized* version of C by considering

$$\mathbb{Q}^\zeta(C(\theta_e, \mathbf{b}) - \mathbf{b} \in \cdot),$$

which, by Lemma 7.5, equals

$$\frac{1}{(d+1)!} \mathbb{E} \int \mathbf{1}\{\operatorname{conv}\{x_0, \ldots, x_d\} - z(x) \in \cdot\} \mathbf{1}_B(z(x)) e^{-\eta(B^d(x_0, \ldots, x_d))} \eta^{d+1}(dx).$$

Remark 7.6 (difference to the graph setting). We note that a similar centralization within O is *not* possible in the setting of typical clusters in a quasi-transitive graph - there is simply no natural way to map one orbit representative to another one.

Further simplifications are possible if the directing random measure η of the regular Cox process ξ is of the special form

$$\eta(\omega, \cdot) = \int \mathbf{1}\{s \in \cdot\} f(\omega, s) \lambda^d(ds), \quad \omega \in \Omega, \tag{7.11}$$

where $f : \Omega \times \mathbb{R}^d \to [0, \infty)$ is measurable, jointly L-invariant and such that ξ is regular. We may then invoke a spherical Blaschke-Petkantschin type formula. To state this formula, we denote by σ the unique $SO(d)$-invariant measure on the sphere

$$S^{d-1} = \{x \in \mathbb{R}^d : ||x|| = 1\},$$

with total mass $\sigma(S^{d-1}) = d \cdot \kappa_d$, where $\kappa_d = \lambda^d(B^d)$. Further, given points $x_0, \ldots, x_d \in \mathbb{R}^d$, we denote by

$$\Delta_d(x_0, \ldots, x_d)$$

the d-dimensional volume of the convex hull of these points. A proof of the following theorem may be found in [64, Theorem 7.3.1] (the proof given there goes back to [54]). The result appeared first in [52] and a different proof than the one presented in [64] may be found in Affentranger [2].

Theorem 7.7 (spherical Blaschke-Petkantschin type formula). *If $f : (\mathbb{R}^d)^{d+1} \to \mathbb{R}$ is a nonnegative measurable function then*

$$\int_{(\mathbb{R}^d)^{d+1}} f(x_0, \ldots, x_d)(\lambda^d)^{d+1}(d(x_0, \ldots x_d))$$
$$= d! \int_{\mathbb{R}^d} \int_0^\infty \int_{S^{d-1}} \cdots \int_{S^{d-1}} f(z + ru_0, \ldots z + ru_d)$$
$$\times r^{d^2-1} \Delta_d(u_0, \ldots, u_d) \sigma(du_0) \ldots \sigma(du_d) dr \lambda^d(dz).$$

As announced, the result in Lemma 7.5 further simplifies if η is of the form (7.11). In the following result, which extends a classical result on typical cells of homogeneous Poisson-Delaunay tessellation due to Miles [52], we fix a specific center function, namely the function z, that assigns to a simplex $K = \text{conv}\{x_0, \ldots, x_d\}$ the center of the ball $B(x_0, \ldots, x_d)$ through its vertices. We call this center function *ball center function*.

Theorem 7.8. (Cox-Delaunay cells for absolutely continuous η) *Let in the setting of Lemma 7.5 the random measure η be of the form (7.11) for a jointly L-invariant measurable $f : \Omega \times \mathbb{R}^d \to [0, \infty)$, then*

$$\mathbb{Q}^\varsigma((C(\theta_e, \mathbf{b}) - \mathbf{b}, \mathbf{b}) \in \cdot) = \frac{1}{d+1} \int_0^\infty \int_{S^{d-1}} \cdots \int_{S^{d-1}} \psi(r, u_0, \ldots, u_d, \cdot)$$
$$\times r^{d^2-1} \Delta_d(u_0, \ldots, u_d) \sigma(du_0) \ldots \sigma(du_d) dr,$$

where the center process is defined with respect to the ball center function and where for fixed $u_0, \ldots, u_d \in S^{d-1}$ and $r \geq 0$ the measure $\psi(r, u_0, \ldots, u_d, \cdot)$ equals

$$\int_{L^\perp} \mathbf{1}\{(r \cdot \text{conv}\{u_0, \ldots, u_d\}, b) \in \cdot\}$$
$$\mathbb{E}\left[e^{-\eta(B(b,r))} f(\theta_e, b + ru_0) \ldots f(\theta_e, b + ru_d)\right] \lambda_{L^\perp}(db). \tag{7.12}$$

Proof. By Lemma 7.5 and the assumption on η we have

$$\mathbb{Q}^\zeta((C(\theta_e, \mathbf{b}) - \mathbf{b}, \mathbf{b}) \in \cdot) = \frac{1}{(d+1)!}\mathbb{E}\int \mathbf{1}\{(\text{conv}\{x_0, \ldots, x_d\} - z(x), \beta(z(x))) \in \cdot\}\mathbf{1}_B(z(x))$$
$$\times e^{-\eta(B^d(x_0, \ldots, x_d))} f(\theta_e, x_0) \ldots f(\theta_e, x_d)(\lambda^d)^{d+1}(dx).$$

Applying Fubini and Theorem 7.7 yields

$$\mathbb{Q}^\zeta((C(\theta_e, \mathbf{b}) - \mathbf{b}, \mathbf{b}) \in \cdot) = \frac{d!}{(d+1)!}\int_{\mathbb{R}^d}\int_0^\infty\int_{S^{d-1}}\cdots\int_{S^{d-1}}\mathbf{1}\{(r \cdot \text{conv}\{u_0, \ldots, u_d\}, \beta(z)) \in \cdot\}$$
$$\times \mathbf{1}_B(z)\mathbb{E}\left[e^{-\eta(B(z,r))}f(\theta_e, z + ru_0) \ldots f(\theta_e, z + ru_d)\right]$$
$$\times r^{d^2-1}\Delta_d(u_0, \ldots, u_d)\sigma(du_0)\ldots\sigma(du_d)dr\lambda^d(dz).$$

Another application of Fubini yields

$$\mathbb{Q}^\zeta((C(\theta_e, \mathbf{b}) - \mathbf{b}, \mathbf{b}) \in \cdot) = \frac{1}{d+1}\int_0^\infty\int_{S^{d-1}}\cdots\int_{S^{d-1}}\psi(r, u_0, \ldots, u_d, \cdot)$$
$$\times r^{d^2-1}\Delta_d(u_0, \ldots, u_d)\sigma(du_0)\ldots\sigma(du_d)dr,$$

where we have written $\psi(r, u_0, \ldots, u_d, \cdot)$ for

$$\int_{\mathbb{R}^d}\mathbf{1}_B(z)\mathbf{1}\{(r \cdot \text{conv}\{u_0, \ldots, u_d\}, \beta(z)) \in \cdot\}$$
$$\mathbb{E}\left[e^{-\eta(B(z,r))}f(\theta_e, z + ru_0) \ldots f(\theta_e, z + ru_d)\right]\lambda^d(dz).$$

Using the orbital decomposition $\lambda^d(dz) = \mu_b(dz)\lambda_{L^\perp}(db)$ of λ with respect to $O = L^\perp$, as well as L-invariance of

$$z \mapsto \mathbb{E}\left[e^{-\eta(B(z,r))}f(\theta_e, z + ru_0) \ldots f(\theta_e, z + ru_d)\right],$$

(which follows by joint L-invariance of f, L-invariance of η and L-invariance of $\mathbb{P}$) the measure $\psi(r, u_0, \ldots, u_d, \cdot)$ reduces to its form given in (7.12). $\qquad\square$

In order to derive some probabilistic interpretations from Lemma 7.5 (or Theorem 7.8), we consider particular jointly L-invariant subsets I of $\Omega \times \mathbb{R}^d$, namely those that are of the special form

$$I = I(\tilde{A}) = \{(\omega, s) : \emptyset \neq C(\omega, s) \in \tilde{A}\}, \tag{7.13}$$

where $\tilde{A} \subset \Delta^{(d)}$ is L-invariant with respect to the induced shifts on $\Delta^{(d)}$. The important property that jointly L-invariant subsets of $\Omega \times \mathbb{R}^d$ of this particular form have, is that $\mathbf{1}_I$ is essentially a function of $C(\omega, s)$, rather than of the whole configuration ω and s. Three typical examples are, given an L-invariant set $A \subset \mathbb{R}^d$, the sets

$$\tilde{A}_1 = \{D \in \Delta^{(d)} : D \subset A\}, \tag{7.14}$$
$$\tilde{A}_2 = \{D \in \Delta^{(d)} : \pi(D) \in A\}, \tag{7.15}$$
$$\tilde{A}_3 = \{D \in \Delta^{(d)} : D \cap A \neq \emptyset\}. \tag{7.16}$$

For the following notion the reader should recall Definition 4.23 and that for any jointly L-invariant measurable subset I of $\Omega \times \mathbb{R}^d$ we have by (4.23)

$$\mathbb{Q}^\zeta(I) = (\mathbb{E}\zeta_I)^*(L^\perp).$$

Definition 7.9 (typical I-cells). Let I denote a jointly L-invariant measurable subset of $\Omega \times \mathbb{R}^d$ and ζ a process of the centers of the cells, defined as in (7.10). If $0 < \mathbb{Q}^\zeta(I) < \infty$, we call

$$\mathbb{P}_\zeta^I(C(\theta_e, \mathbf{b}) - \mathbf{b} \in \cdot) = \frac{1}{(\mathbb{E}\zeta_I)^*(L^\perp)} \int \mathbf{1}\{C(\theta_e, b) - b \in \cdot\} \mathbf{1}_I(\omega, b) \mathbb{Q}^\zeta(d(\omega, b))$$

the *distribution of the I-typical cell*. If for a given L-invariant measurable subset A of $\mathbb{R}^d$ the set I is of the special form

(i) $\{(\omega, s) : \emptyset \neq C(\omega, s) \in \tilde{A}_1\}$,

(ii) $\{(\omega, s) : \emptyset \neq C(\omega, s) \in \tilde{A}_2\}$,

(iii) $\{(\omega, s) : \emptyset \neq C(\omega, s) \in \tilde{A}_3\}$,

we call $\mathbb{P}_\zeta^I(C(\theta_e, \mathbf{b}) - \mathbf{b} \in \cdot)$ the distribution of the *typical cell contained in A*, the distribution of the *typical cell with center in A* and the distribution of the *typical cell intersecting A*, respectively.

Note that, as explained in the introduction of Chapter 4, the word *typical* has to be read with care in this generality. We shall only consider probabilistic interpretations of Theorem 7.8 (Lemma 7.5 may be used analogously under the corresponding weaker conditions), where we restricted our attention to the ball center function.

Corollary 7.10 (probabilistic interpretations). *Let $\tilde{A} \subset \Delta^{(d+1)}$ be measurable and L-invariant and $I = I(\tilde{A})$ be defined as in (7.13). In the situation of Lemma 7.8 the typical I-cell has distribution*

$$\frac{1}{(\mathbb{E}\zeta_I)^*(L^\perp)} \frac{1}{d+1} \int_0^\infty \int_{S^{d-1}} \cdots \int_{S^{d-1}} \mathbf{1}\{r \cdot \mathrm{conv}\{u_0, \ldots, u_d\} \in \cdot\}$$
$$\times \psi(r, u_0, \ldots, u_d) r^{d^2-1} \Delta_d(u_0, \ldots, u_d) \sigma(du_0) \ldots \sigma(du_d) dr,$$

where $\psi(r, u_0, \ldots, u_d)$ is given by

$$\int_{L^\perp} \mathbf{1}\{r \cdot \mathrm{conv}\{u_0, \ldots, u_d\} + b \in \tilde{A}\} \mathbb{E}\left[e^{-\eta(B(b,r))} f(\theta_e, b + ru_0) \ldots f(\theta_e, b + ru_d)\right] \lambda_{L^\perp}(db).$$

Proof. By definition we have for any jointly L-invariant $\tilde{A} \subset \Delta^{(d+1)}$

$$\mathbb{P}_\zeta^I(C(\theta_e, \mathbf{b}) - \mathbf{b} \in \cdot) = \frac{1}{(\mathbb{E}\zeta_I)^*(L^\perp)} \int \mathbf{1}\{C(\theta_e, b) - b \in \cdot, \emptyset \neq C(\theta_e, b) \in \tilde{A}\} \mathbb{Q}^\zeta(d(\omega, b)),$$

and by Theorem 7.8 the right hand side equals

$$\frac{1}{(\mathbb{E}\zeta_I)^*(L^\perp)} \frac{1}{d+1} \int_0^\infty \int_{S^{d-1}} \cdots \int_{S^{d-1}} \int_{L^\perp} \mathbf{1}\{r \cdot \mathrm{conv}\{u_0, \ldots, u_d\} \in \cdot\}$$
$$\times \mathbf{1}\{r \cdot \mathrm{conv}\{u_0, \ldots, u_d\} + b \in \tilde{A}\} \mathbb{E}\left[e^{-\eta(B(b,r))} f(\theta_e, b + ru_0) \ldots f(\theta_e, b + ru_d)\right]$$
$$\times r^{d^2-1} \Delta_d(u_0, \ldots, u_d) \lambda_{L^\perp}(db) \sigma(du_0) \ldots \sigma(du_d) dr,$$

This yields the assertion. □

Remark 7.11. This corollary extends known results from the special case of a completely stationary Poisson-Delaunay mosaic, established in [52] and [53], apart from the explicit computation of the constant

$$
(\mathbb{E}\zeta_I)^*(L^\perp) = \frac{1}{d+1} \int_0^\infty \int_{S^{d-1}} \cdots \int_{S^{d-1}} \int_{L^\perp} \mathbf{1}\{r \cdot \mathrm{conv}\{u_0, \ldots, u_d\} + b \in \tilde{A}\}
$$
$$
\times \mathbb{E}\left[e^{-\eta(B(b,r))} f(\theta_e, b + ru_0) \ldots f(\theta_e, b + ru_d)\right]
$$
$$
\times r^{d^2-1} \Delta_d(u_0, \ldots, u_d) \lambda_{L^\perp}(db) \sigma(du_0) \ldots \sigma(du_d) dr.
$$

A more detailed analysis of the typical (A-)cell of a Cox Delaunay mosaic and its k-faces along the lines of Baumstark and Last [5] seems feasible for (even partially) stationary Cox processes.

7.2 Random partitions

The object of interest in this section are random partitions of some topological space S that are stationary with respect to an operating group G. Random partitions were first introduced and inspected by Last [36] in full generality and we shall follow his approach here. To define a G-stationary random partition, let the lcsc group G operate on the measurable space S properly. Given a G-stationary simple point process ξ on S we may define a G-stationary partition based on ξ as a measurable mapping $\pi : \Omega \times S \to S$ that satisfies

$$
\pi(\omega, s) \in \xi(\omega), \quad s \in S, \xi(\omega) \neq \emptyset,
$$
$$
\pi(\omega, s) = s, \quad s \in S, \xi(\omega) = \emptyset,
$$

and is G-covariant in the sense that

$$
\pi(\theta_g \omega, gs) = g\pi(\omega, s), \quad \omega \in \Omega, s \in S, g \in G.
$$

We define

$$
C^\pi(\omega, s) := \{t \in S : \pi(\omega, t) = s\}, \quad \omega \in \Omega, s \in S,
$$

and note that $C^\pi(\omega, s) = \emptyset$ whenever $s \notin \xi(\omega) \neq \emptyset$. Note that for $\omega \in \{\xi = \emptyset\}$ we have $C^\pi(\omega, s) = s, s \in S$. In addition, we remark that $\pi(\omega, s)$ needs not be an element of $C^\pi(\omega, s)$ and that

$$
\bigcup_{s \in \xi(\omega)} C^\pi(\omega, s) = S, \quad \xi(\omega) \neq \emptyset.
$$

We refer for $s \in \xi(\omega)$ to $C^\pi(\omega, s)$ as the *cell* with *center* s in configuration ω. Some of these cells may very well be empty. In addition, we refer to $\pi(\omega, s)$ as the *center of s in configuration* ω and then

$$
V^\pi(\omega, s) := C^\pi(\omega, \pi(\omega, s)), \quad \omega \in \Omega, \tag{7.17}
$$

is the cell containing $s \in S$. G-covariance of π readily implies G-covariance of C^π and V^π in the sense that both

$$C^\pi(\theta_g\omega, gs) = gC^\pi(\omega, s), \quad g \in G, \omega \in \Omega, s \in S,$$

and

$$V^\pi(\theta_g\omega, gs) = gV^\pi(\omega, s), \quad g \in G, \omega \in \Omega, s \in S.$$

We want to enable us to speak of the distribution of $C^\pi(\mathbf{b})$ under the cumulative Palm measure of the center process and of the distribution of $V^\pi(t)$ for $t \in S$. In order to avoid larger technical issues we restrict ourselves now to a topological setting, where S is a lcsc topological space. In this case, we may equip the space of all closed subsets of S with the topology of closed convergence (also called *Fell topology*), see [64, 28]. Instead of looking at $C^\pi(\omega, s), s \in S, \omega \in \Omega$, or $V^\pi(\omega, s), s \in S, \omega \in \Omega$, we may replace them by their closures in S, denoted by $\bar{C}^\pi(\omega, s), s \in S, \omega \in \Omega$, and $\bar{V}^\pi(\omega, s), s \in S, \omega \in \Omega$. It is then enough to require that

$$(\omega, s) \mapsto \bar{C}^\pi(\omega, s),$$

is measurable with respect to $\mathcal{A} \otimes \mathcal{S}$ and the Borel σ-field induced by the Fell topology on the space of closed subsets of S. A random partition π based on a simple point process ξ on S satisfying this requirement will be called a *random topological partition*. If ξ is G-stationary, we call π G-stationary as well. The corresponding measurability of

$$(\omega, s) \mapsto \bar{V}^\pi(\omega, s),$$

for a random topological partition then clearly follows from (7.17).

In Subsection 7.2.1 we relate the quasi-distributions of suitably defined ν-weighted 0-cells and typical cells for G-stationary random topological partitions, where ν is an arbitrary G-invariant measure on the topological space S. We then apply these results in Subsection 7.2.2 to the setting of orientable Riemannian manifolds, where the operating group is the isometry group and ν is the isometry invariant surface measure, after briefly summarizing the most important definitions. We then conclude the section by illustrating the theory with some examples in Subsection 7.2.3.

7.2.1   Relations between typical and 0-cells

Our task for this subsection will be to suitably define 0-cells and typical cells of a G-stationary random topological partition and to relate their distributions. Note at this point that we neither require transitivity of the operation nor unimodularity of the operating group. Given a measure ν on S, we write (similar to a notion in Timar [67])

$$\nu^\Delta(\cdot) = \int \mathbf{1}\{s \in \cdot\}\Delta^*(s)\nu(ds) \tag{7.18}$$

and call it the Δ-*weighted ν*. We start with the following consequence of the Palm MTP, which in parts extends results of Last [38] to a non-transitive setting.

Theorem 7.12 (typical cells and 0-cells). *Let G operate properly on the lcsc space S and let π denote a G-stationary random topological partition based on the simple point process ξ. Let further ν denote a G-invariant σ-finite measure on S. Then*

$$\int \mathbb{E} \int \mathbf{1}\{g^{-1}\bar{V}^\pi(b) \in \cdot\}\kappa_{\beta(\pi(b)),\pi(b)}(dg)\nu^*(db)$$
$$= \int \mathbf{1}\{\bar{C}^\pi(\omega,b) \in \cdot\}\nu^\Delta(C^\pi(\omega,b))\mathbb{Q}^\xi(d(\omega,b)), \qquad (7.19)$$

and if π is such that $0 < \nu(\bar{C}^\pi(\omega,s)) < \infty, s \in \xi(\omega), \omega \in \Omega$, then

$$\int \mathbb{E} \int \Delta^*(\pi(b))\mathbf{1}\{g^{-1}\bar{V}^\pi(b) \in \cdot\}\kappa_{\beta(\pi(b)),\pi(b)}(dg)\frac{1}{\nu(V^\pi(b))}\nu^*(db)$$
$$= \int \mathbf{1}\{\bar{C}^\pi(\omega,b) \in \cdot\}\mathbb{Q}^\xi(d(\omega,b)). \qquad (7.20)$$

Proof. Putting in the Palm MTP (Theorem 5.6) $\eta(\omega) := \nu$, $\gamma(\omega,s,dt) := \mathbf{1}\{t \in C^\pi(\omega,s)\}\nu(dt)$ and $\delta(\omega,t,ds) := \delta_{\pi(\omega,t)}(ds)$ we note that

$$\iint \mathbf{1}\{(s,t) \in \cdot\}\gamma(s,dt)\xi(ds) = \iint \mathbf{1}\{(s,t) \in \cdot\}\mathbf{1}\{t \in C^\pi(\omega,s)\}\nu(dt)\xi(ds)$$
$$= \iint \mathbf{1}\{(\pi(t),t) \in \cdot\}\mathbf{1}\{t \in C^\pi(\omega,s)\}\nu(dt)\xi(ds)$$

and that, using Fubini, the right-hand side reduces to

$$\int \mathbf{1}\{(\pi(\omega,t),t) \in \cdot\}\nu(dt) = \iint \mathbf{1}\{(s,t) \in \cdot\}\delta(t,ds)\nu(dt).$$

Thus Theorem 5.6 yields

$$\iint \Delta^*(t)m(\omega,b,t)\mathbf{1}\{t \in C^\pi(\omega,b)\}\nu(dt)\mathbb{Q}^\xi(d(\omega,b))$$
$$= \int m(\omega,\pi(\omega,b),b)\mathbb{Q}^\eta(d(\omega,b)) \qquad (7.21)$$

for arbitrary jointly G-invariant $m : \Omega \times S \times S \to [0,\infty)$. Putting

$$m(\omega,s,t) := \int \mathbf{1}\{g^{-1}\bar{C}^\pi(\omega,s) \in D\}\kappa_{\beta(s),s}(dg)$$

for an arbitrary, but fixed measurable subset D of the space of closed subsets of S, the left-hand side of (7.21) may be written, using Fubini, $G_{b,b}$-invariance of Δ^* and G-covariance of C^π and $\bar{C}^\pi$, as

$$\iiint \mathbf{1}\{\bar{C}^\pi(\theta_g^{-1}\omega,b) \in \cdot\}\Delta^*(g^{-1}t)\mathbf{1}\{g^{-1}t \in C^\pi(\theta_g^{-1}\omega,b)\}\nu(dt)\kappa_{b,b}(dg)\mathbb{Q}^\xi(d(\omega,b)),$$

where we omitted D. Using G-invariance of ν and then (4.3), this reduces to

$$\int \mathbf{1}\{\bar{C}^\pi(\omega,b) \in \cdot\}\nu^\Delta(C^\pi(\omega,b))\mathbb{Q}^\xi(d(\omega,b)).$$

On the other hand, the right-hand side of (7.21) may be written by (4.7) as

$$\int \mathbb{E} \int \mathbf{1}\{g^{-1}\bar{C}^\pi(\pi(b)) \in \cdot\}\kappa_{\beta(\pi(b)),\pi(b)}(dg)\nu^*(db).$$

This yields (7.19) as $\bar{C}^\pi(\pi(s)) = \bar{V}^\pi(s), s \in S$, ω-wise. Equation (7.20) follows from using in (7.21)

$$m(\omega,s,t) = \tilde{\Delta}(t,s)\int \mathbf{1}\{g^{-1}\bar{C}^\pi(\omega,s) \in \cdot\}\kappa_{\beta(s),s}(dg)\frac{1}{\nu(C^\pi(\omega,s))}$$

and a similar calculation. $\qquad\square$

7.2.2 Riemannian manifolds

Theorem 7.12 applies in particular to the setting where S is a k-dimensional orientable Riemannian manifold, G is its isometry group $I(S)$ endowed with the compact-open topology (see below) and ν is any $I(S)$-invariant measure on S, e.g. its surface measure, which we denote by μ_S. An introduction to the subject as well as definitions of the above notions may be found in [41, 40], and a discussion of topological aspects on the isometry group of a Riemannian manifold may be found in [25]. Here are some more detailed definitions.

If (S, g_S) and (T, g_T) are Riemannian manifolds with metric tensors g_S and g_T respectively, then an *isometry* between S and T is a diffeomorphism $f : S \to T$ respecting the given metrics in the sense that

$$\langle u, v \rangle_p = \langle df_p(u), df_p(v) \rangle_{f(p)}, \quad u, v \in T_p S, p \in S,$$

where $T_p S$ denotes the tangent space of S in the point $p \in S$, $df_p : T_p S \to T_{f(p)} T$ the differential of f in the point $p \in S$ and $\langle \cdot, \cdot \rangle_p$ and $\langle \cdot, \cdot \rangle_{f(p)}$ the inner products in p and $f(p)$ induced by the respective metric tensors. A consequence of the definition is that isometries indeed preserve distances, i.e.

$$d_S(p, q) = d_T(f(p), f(q)), \quad p, q \in S, \tag{7.22}$$

where we recall that the *distance function* $d_S(p, q)$ is defined as the infimum over the lengths of all paths in S connecting $p \in S$ and $q \in S$. In fact, given a map $f : S \to S$ satisfying (7.22) it can be shown that f is already an isometry in the sense of our above definition, cf. [25, Theorem 11.1]. If the manifold S is orientable, there is a unique *Riemannian volume form* [41], which induces a measure μ_S on the Borel σ-algebra $\mathcal{B}(S)$ on S which we will call *surface measure* of S. Another important consequence is, that the surface measure μ_S of an orientable manifold S is invariant with respect to $G = I(S)$. The set of isometries on S forms the *isometry group* $I(S)$. It is given the *compact-open* topology generated by all sets of the form

$$W(C, U) := \{g \in I(S) : g(C) \subset U\}$$

where $C \subset S$ is compact and $U \subset S$ is open. With respect to this topology $I(S)$ becomes a locally compact second countable Hausdorff topological group and the operation of $I(S)$ on S is continous, cf. [25]. Also, $I(S) \hookrightarrow S$ is topologically proper as the following lemma shows.

Lemma 7.13 (isometries act proper). *The operation of $I(S)$ on S is topologically proper.*

Proof. Let $K \subset S$ be compact and let f_n denote a sequence in $\pi_s^{-1}(K)$. Then $f_n(s)$ is a sequence in K and by compactness of K there is a subsequence $f_{h(n)}$ of f_n with $f_{h(n)}(s) \to t$ for some fixed $t \in K$. Now [25, Theorem 2.2] states that there is a further subsequence $f_{h'(n)}$ and some $f \in I(S)$ such that $f_{h'(n)}$ converges to f in the compact-open topology. This shows that $\pi_s^{-1}(K)$ is sequentially compact. Since $I(S)$ is second countable it follows that $\pi_s^{-1}(K)$ is in fact compact. $\qquad\square$

It is an important fact that isometry groups of compact manifolds are compact themselves (and in particular unimodular).

Lemma 7.14 (compactness). *The isometry group $I(S)$ of a compact Riemannian manifold is compact.*

Proof. Let f_n denote a sequence in $I(S)$ and $s \in S$ some fixed point. Then $f_n(s)$ is a sequence in S and by compactness of S there is a subsequence $f_{h(n)}$ of f_n with $f_{h(n)}(s) \to t$ for some fixed $t \in S$. By [25, Theorem 2.2] there is a further subsequence $f_{h'(n)}$ and some $f \in I(S)$ such that $f_{h'(n)}$ converges to f in the compact-open topology. This shows that $I(S)$ is sequentially compact. Since $I(S)$ is second-countable it follows that $I(S)$ is in fact compact. $\qquad\square$

We may use the cumulative Palm measure to define the following random object.

Definition 7.15 (typical cells). Let S denote an orientable Riemannian manifold with surface measure μ_S and let O denote a measurable system of orbit representatives. Let further π denote an $I(S)$-stationary random topological tessellation on S based on the simple point process ξ. Let $\tilde{A}$ denote an $I(S)$-invariant measurable subset of the space of closed subsets of S, define the jointly $I(S)$-invariant set

$$I := \{(\omega, s) \in \Omega \times S : C^{\pi}(\omega, s) \in \tilde{A}\},$$

and let $\tilde{A}$ be such that $0 < (\mathbb{E}\xi_I)^*(S) < \infty$. Then a random closed subset of S with distribution

$$\frac{1}{(\mathbb{E}\xi_I)^*(S)} \int \mathbf{1}\{\bar{C}^{\pi}(\omega, b) \in \cdot\}\mathbf{1}\{\bar{C}^{\pi}(\omega, b) \in \tilde{A}\}\mathbb{Q}^{\xi}(d(\omega, b))$$

is called a *typical cell* of π under all cells with the property $\tilde{A}$, or in short a *typical $\tilde{A}$-cell*.

Let π denote an $I(S)$-stationary random topological tessellation on the Riemannian manifold S based on the simple point process ξ, and let O denote a measurable system of orbit representatives. To omit difficulties arising from boundary effects when using such $I(S)$-invariant sets $\tilde{A}$, we now restrict our attention to the case, when we may choose $\tilde{A}$ to be the complete space of closed subsets of S, i.e. to the case when both $0 < (\mathbb{E}\xi)^*(S) < \infty$ and $0 < \mu_S^*(S) < \infty$. In this case, the *typical cell* of π is defined as a random closed set with distribution

$$\int \mathbf{1}\{\bar{C}^{\pi}(\omega, b) \in \cdot\}\mathbb{P}^{\xi}(d(\omega, b)),$$

where we recall that

$$\mathbb{P}^{\xi} = \frac{1}{\mathbb{Q}^{\xi}(\Omega \times S)}\mathbb{Q}^{\xi} = \frac{1}{(\mathbb{E}\xi)^*(S)}\mathbb{Q}^{\xi},$$

while a *volume weighted typical cell* of π is a random compact set with distribution

$$\frac{1}{\mathbb{E}^{\xi}\mu_S(C^{\pi})} \int \mathbf{1}\{\bar{C}^{\pi}(\omega, b) \in \cdot\}\mu_S(C^{\pi}(\omega, b))\mathbb{P}^{\xi}(d(\omega, b)),$$

whenever $0 < \mathbb{E}^{\xi}\mu_S(C^{\pi}) < \infty$. Similarly, we define the Δ-*volume weighted typical cell* as a random closed set with distribution

$$\frac{1}{\mathbb{E}^{\xi}\mu_S^{\Delta}(C^{\pi})} \int \mathbf{1}\{\bar{C}^{\pi}(\omega, b) \in \cdot\}\mu_S^{\Delta}(C^{\pi}(\omega, b))\mathbb{P}^{\xi}(d(\omega, b)),$$

whenever $0 < \mathbb{E}^{\xi} \mu_S^{\Delta}(C^{\pi}) < \infty$. Further, we call a random closed subset of S with distribution

$$\frac{1}{\mu_S^*(S)} \int \mathbb{E} \int \mathbf{1}\{g^{-1}\bar{V}^{\pi}(b) \in \cdot\} \kappa_{\beta(\pi(b)),\pi(b)}(dg) \mu_S^*(db)$$

a *centralized 0-cell*, while a random closed subset of S with distribution

$$\frac{1}{C} \int \mathbb{E} \int \Delta^*(\pi(b)) \mathbf{1}\{g^{-1}\bar{V}^{\pi}(b) \in \cdot\} \kappa_{\beta(\pi(b)),\pi(b)}(dg) \frac{1}{\mu_S(V^{\pi}(b))} \mu_S^*(db)$$

where $C := \int \mathbb{E} \left[\Delta^*(\pi(b)) \frac{1}{\mu_S(V^{\pi}(b))} \right] \mu_S^*(db)$ is called a *centralized Δ-picked volume debiased 0-cell*.

What follows is a probabilistic interpretation of Theorem 7.12 in this manifold setting.

Corollary 7.16 (probabilistic interpretations). *Let S denote an orientable Riemannian manifold and G a closed subgroup of $I(S)$. Let further π denote a G-stationary random topological partition of S based on the simple point process ξ such that $0 < (\mathbb{E}\xi)^*(S) < \infty$. Let N denote a centralized 0-cell, n_w^{Δ} a Δ-weighted-picked centralized and size-debiased 0-cell, Z a typical cell and Z_s^{Δ} a Δ-volume weighted typical cell. Then their distributions are related via*

$$\mathbb{P}(N \in \cdot) = \mathbb{P}(Z_s^{\Delta} \in \cdot), \tag{7.23}$$

$$\mathbb{P}(n_w^{\Delta} \in \cdot) = \mathbb{P}(Z \in \cdot), \tag{7.24}$$

and we have the relations

$$\mu_S^*(S) = \mathbb{E}^{\xi} \left[\mu_S^{\Delta}(C^{\pi}(\theta_e, \mathbf{b})) \right] (\mathbb{E}\xi)^*(S) \tag{7.25}$$

$$\int \mathbb{E} \left[\Delta^*(\pi(b)) \frac{1}{\mu_S(V^{\pi}(b))} \right] \mu_S^*(db) = (\mathbb{E}\xi)^*(S). \tag{7.26}$$

Proof. Rewriting all sides of the equations in Theorem 7.12 in terms of the distributions introduced above the corollary yields the equations

$$\mu_S^*(S)\mathbb{P}(N \in \cdot) = \mathbb{E}^{\xi} \left[\mu_S^{\Delta}(C^{\pi}(\theta_e, \mathbf{b})) \right] (\mathbb{E}\xi)^*(S)\mathbb{P}(Z_S^{\Delta} \in \cdot)$$

$$\int \mathbb{E} \left[\Delta^*(\pi(b)) \frac{1}{\mu_S^*(V^{\pi}(b))} \right] \mu_S^*(db)\mathbb{P}(n_w^{\Delta} \in \cdot) = (\mathbb{E}\xi)^*(S)\mathbb{P}(Z \in \cdot)$$

Plugging in the space of all closed subsets of S on either side yields the assertions. $\square$

7.2.3   Examples

In this subsection we illustrate Corollary 7.16 by giving some examples.

Example 7.17 ($\mathbb{Z}^d \hookrightarrow \mathbb{R}^d$). As $\mathbb{Z}^d$ is unimodular, the modular function, being identically 1 (also compare Lemma 3.17), may be omitted in all formulas. We have $\mu_{\mathbb{R}^d} = \lambda^d$ and

$$\mu_{\mathbb{R}^d}^* = (\lambda^d)^* = \lambda^d(\cdot \cap [0,1)^d),$$

and the inversion kernel is given by

$$\kappa_{\beta(s),s}(\cdot) = \delta_{s-\beta(s)}(\cdot) = \delta_{\lfloor s \rfloor}(\cdot), \quad s \in \mathbb{R}^d,$$

where $\lfloor s \rfloor := (\lfloor s_1 \rfloor, \ldots, \lfloor s_d \rfloor)$ is the component wise Gauss bracket, assigning to a real number its integer part. The centralized 0-cell has then by definition the distribution

$$\int \mathbb{P}(\bar{V}^\pi(b) - \lfloor \pi(b) \rfloor \in \cdot) \mathbf{1}_{[0,1)^d}(b) \lambda^d(db),$$

which is obviously the distribution of a centralized cell, picked by realizing the random partition π, independently realizing a uniformly distributed random vector U in $O = [0,1)^d$ and then forming the closure of the unique cell $V^\pi(U)$ such that $U \in V^\pi(U)$. Thus (7.23) reduces to the equality

$$V^\pi(U) \stackrel{d}{=} Z_s,$$

where Z_s is a version of the ordinary volume weighted typical cell. Equation (7.24) simplifies similarly and tells us that the typical cell may be interpreted as a size-debiased version of the 0-cell. The relations (7.25), (7.26) reduce to

$$1 = \mathbb{E}^\xi \left[\lambda^d(C^\pi(\theta_e, \mathbf{b})) \right] \mathbb{E}\xi([0,1)^d)$$

$$\int \mathbb{E}\left[\frac{1}{\lambda^d(V^\pi(b))} \right] \mathbf{1}_{[0,1)^d}(b) \lambda^d(db) = \mathbb{E}\xi([0,1)^d).$$

In particular we have

$$\mathbb{E}\left[\int \frac{1}{\lambda^d(V^\pi(b))} \mathbf{1}_{[0,1)^d}(b) \lambda^d(db) \right] = \frac{1}{\mathbb{E}^\xi \left[\lambda^d(C^\pi(\theta_e, \mathbf{b})) \right]} = \mathbb{E}\xi([0,1)^d).$$

Example 7.18 (infinite cylinder). We consider the non-transitive operation of $\mathbb{R}$ on the infinite cylinder $C := \mathbb{R} \times S^1$ by shifting the first coordinate. Here S^1 denotes the one-dimensional unit circle in $\mathbb{R}^2$. Again, since $\mathbb{R}$ is unimodular, the modular function vanishes in all formulas. We have for arbitrary measurable $f : C \to [0, \infty]$ the equality

$$\mu_C f = \iint f(g + b) \lambda^1(dg) \lambda_{S^1}(db),$$

and choosing λ^1 as Haar measure on $\mathbb{R}$ and $O := \{0\} \times S^1$, we obtain $\mu_C^* = \lambda_{S^1}$. The centralized 0-cell has by definition the distribution

$$\frac{1}{2\pi} \int \mathbb{E} \int \mathbf{1}\{\bar{V}^\pi(b) - x \in \cdot\} \kappa_{\beta(\pi(b)),\pi(b)}(dx) \lambda_{S^1}(db),$$

which is obviously the distribution of a centralized cell, picked by realizing the random partition π, independently realizing a uniformly distributed random vector U in $O = \{0\} \times S^1$, and then forming the closure of the unique cell $V^\pi(U)$ such that $U \in V^\pi(U)$ (see Figure 7.2). Thus (7.23) reduces to the equality

$$V^\pi(U) - \pi(U) \stackrel{d}{=} Z_s,$$

where Z_s is a version of the ordinary volume weighted typical cell. Equation (7.24)

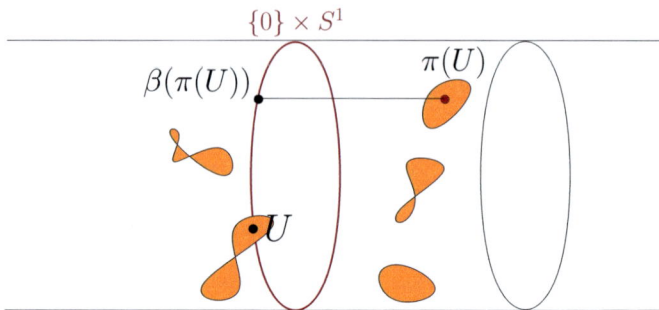

Figure 7.1: Realization of a 0-cell (orange region)

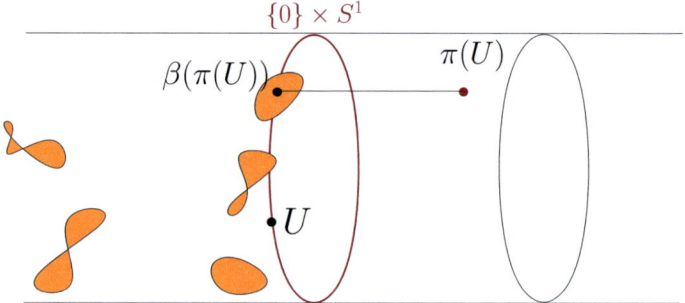

Figure 7.2: Centralized version of the above 0-cell

simplifies similarly and tells us that the typical cell may be interpreted as a size-debiased version of the 0-cell. The relations (7.25), (7.26) reduce to

$$2\pi = \mathbb{E}^{\xi}\left[\mu_C(C^{\pi}(\theta_e, \mathbf{b}))\right](\mathbb{E}\xi)^*(\{0\} \times S^1)$$

$$\int \mathbb{E}\left[\frac{1}{\mu_C(V^{\pi}(b))}\right]\lambda_{S^1}(db) = (\mathbb{E}\xi)^*(\{0\} \times S^1).$$

In particular, we have

$$\mathbb{E}\left[\int \frac{1}{\mu_C(V^{\pi}(b))}\lambda_{S^1}(db)\right] = \frac{2\pi}{\mathbb{E}^{\xi}\left[\mu_C(C^{\pi}(\theta_e, \mathbf{b}))\right]} = (\mathbb{E}\xi)^*(\{0\} \times S^1).$$

Example 7.19 (hyperbolic plane)**.** We give a transitive, but non-unimodular example here. We consider the upper half-plane

$$\mathbb{H}^2 := \{(x, y) \in \mathbb{R}^2 : y > 0\} \simeq \{z \in \mathbb{C} : \operatorname{Im}(z) > 0\},$$

and the group

$$G := \left\{\begin{pmatrix} y & x \\ 0 & 1 \end{pmatrix} : y > 0, x \in \mathbb{R}\right\},$$

with respect to the usual multiplication of matrices, endowed with the metric topology of $\mathbb{R} \times \mathbb{R}_{>0}$ with the obvious identification, which is the same as the inherited topology from $\mathbb{R}^4$. We let G act on $\mathbb{H}^2$ (identified with the complex upper half-plane to avoid conflicts with the notation for matrix-vector multiplication) via

$$\begin{pmatrix} y & x \\ 0 & 1 \end{pmatrix} \cdot z := yz + x, \quad z \in \mathbb{H}^2, x \in \mathbb{R}, y > 0.$$

It is straightforward to convince oneself that this gives an operation indeed, and we shall denote it by $G \hookrightarrow \mathbb{H}^2$. We now determine a left Haar measure on G (also see [19, p. 359]). Identifying the matrix

$$\begin{pmatrix} y & x \\ 0 & 1 \end{pmatrix}$$

with the pair $(x, y) \in \mathbb{H}^2$, we may define the measure

$$\lambda(f) := \int_G \frac{f(x,y)}{y^2} \lambda^2(d(x,y))$$

on G. It is clearly non-zero and locally finite, and it is left-invariant: Let $(a, b), (x, y) \in G$. Then $(a, b) \cdot (x, y)$ is the product

$$\begin{pmatrix} b & a \\ 0 & 1 \end{pmatrix} \cdot \begin{pmatrix} y & x \\ 0 & 1 \end{pmatrix} = \begin{pmatrix} by & bx + a \\ 0 & 1 \end{pmatrix},$$

and is thus identified with $(bx + a, by) =: T_{(a,b)}(x, y)$, where $T_{(a,b)}$ maps $\mathbb{H}^2$ to itself. Then

$$\int_G \frac{f((a,b) \cdot (x,y))}{y^2} \lambda^2(d(x,y)) = \int_G \frac{f(T_{(a,b)}(x,y))}{b^2 y^2} \cdot b^2 \lambda^2(d(x,y))$$

$$= \int_G g(T_{(a,b)}(x,y)) \cdot b^2 \lambda^2(d(x,y)),$$

where we have put $g(x, y) := f(x, y)/y^2$. Furthermore, the Jacobian of $T_{(a,b)}$ at (x, y) equals

$$\begin{pmatrix} b & 0 \\ 0 & b \end{pmatrix},$$

and thus the absolute value of its determinant equals b^2. Now the transformation theorem gives

$$\int_G \frac{f((a,b) \cdot (x,y))}{y^2} \lambda^2(d(x,y)) = \int_G g(x,y) \lambda^2(d(x,y))$$

$$= \int_G \frac{f(x,y)}{y^2} \lambda^2(d(x,y)),$$

which is the desired left-invariance. Clearly, if λ is a left Haar measure, then

$$\tilde{\lambda} f := \lambda(\tilde{f}), \quad f \in \mathcal{G}_+, \quad \text{where} \quad \tilde{f}(g) = f(g^{-1}), \quad g \in G,$$

defines a right Haar measure. We have $(x, y)^{-1} = (-x/y, 1/y) =: T(x, y)$. The Jacobian of T at (x, y) is given by

$$\begin{pmatrix} -\frac{1}{y} & \frac{x}{y^2} \\ 0 & -\frac{1}{y^2} \end{pmatrix},$$

whose determinant equals $1/y^3$ (which is positive, since $y > 0$). Putting now $g(x, y) := f(x, y)/y$, we have

$$\tilde{\lambda}f = \int_G \frac{f(T(x, y))}{y^2}\lambda^2(d(x, y)) = \int_G g(T(x, y)) \cdot \frac{1}{y^3}\lambda^2(d(x, y))$$
$$= \int_G g(T(x, y))|\det DT(x, y)|\lambda^2(d(x, y)),$$

and thus the transformation theorem yields

$$\tilde{\lambda}f = \int_G g(x, y)\lambda^2(d(x, y)) = \int_G \frac{f(x, y)}{y}\lambda^2(d(x, y)).$$

To compute the modular function, it is enough to note that from (2.2)

$$\int \Delta(g^{-1})f(g)\lambda(dg) = \int f(g^{-1})\lambda(dg) = \int f(g)\tilde{\lambda}(dg),$$

which means

$$\int_G \Delta((x, y)^{-1})\frac{f(x, y)}{y^2}\lambda^2(d(x, y)) = \int \frac{f(x, y)}{y}\lambda^2(d(x, y)).$$

Then a comparison yields $\Delta((x, y)^{-1}) = y$ and thus

$$\Delta((x, y)) = \Delta\left(\begin{pmatrix} y & x \\ 0 & 1 \end{pmatrix}\right) = y^{-1}, \quad x \in \mathbb{R}, y > 0.$$

We fix the imaginary unit i as the 'origin' of the upper half-plane and consider the pushforward of the left Haar measure on G under the projection π_i. Since

$$\pi_i((x, y)) = (x, y) \cdot i = y \cdot i + x,$$

we receive

$$\mu_i f = \lambda(f \circ \pi_i) = \int_G \frac{f(x + y \cdot i)}{y^2}\lambda^2(d(x, y)) = \int_{\mathbb{H}^2} f(z)\mu_\mathbb{H}(dz).$$

Thus μ_i agrees with the well-known hyperbolic Riemannian surface measure $\mu_\mathbb{H}$. Since $G \hookrightarrow \mathbb{H}^2$ is an injective and transitive action, we have

$$\kappa_{i,s} = \delta_{g_{i,s}}, \quad s \in \mathbb{H}^2,$$

where $g_{i,s} = \begin{pmatrix} s_y & s_x \\ 0 & 1 \end{pmatrix}$ is the unique element in G shifting i to $s = (s_x, s_y) \in \mathbb{H}^2$ and thus

$$\Delta^*(s) = \Delta\left(\begin{pmatrix} s_y & s_x \\ 0 & 1 \end{pmatrix}^{-1}\right) = \Delta\left(\begin{pmatrix} 1/s_y & -s_x/s_y \\ 0 & 1 \end{pmatrix}\right) = s_y, \quad s \in \mathbb{H}^2.$$

We conclude, that the Δ-weighted hyperbolic volume measure is given by

$$\mu_i^\Delta(\cdot) = \int \frac{\mathbf{1}\{x + iy \in \cdot\}}{y^2} \cdot y\lambda^2(d(x, y)) = \int \frac{\mathbf{1}\{x + iy \in \cdot\}}{y}\lambda^2(d(x, y)). \qquad (7.27)$$

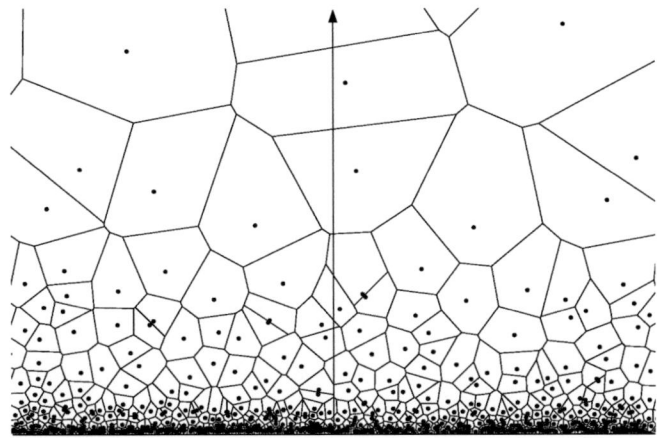

Figure 7.3: Realization of an euclidean Voronoi tessellation based on a homogeneous Poisson process in hyperbolic space

Further, with respect to $O := \{i\}$ we compute

$$\mu^*_{\mathbb{H}^2} = \mu^*_i = \delta_i.$$

Suppose we are given a G-stationary simple point process ξ on $\mathbb{H}^2$ of intensity γ_ξ, along with a G-stationary random topological partition π based on ξ. E.g. ξ might be a Poisson process with intensity measure μ_i, and π such that the closures of the cells coincide with a Voronoi-mosaic either with respect to the euclidean metric (see Figure 7.3), or with respect to the hyperbolic metric. Then by definition, the distribution of the centralized 0-cell (which might be called more accurately *centralized i-cell* in this setting) is given by

$$\mathbb{P}\left(\begin{pmatrix} \pi(i)_y & \pi(i)_x \\ 0 & 1 \end{pmatrix}^{-1} \cdot \bar{V}^\pi(i) \in \cdot\right) = \mathbb{P}\left(\frac{1}{\pi(i)_y}\bar{V}^\pi(i) - \frac{\pi(i)_x}{\pi(i)_y} \in \cdot\right).$$

Furthermore, the distribution of the centralized Δ-picked volume debiased 0-cell is given by

$$\frac{1}{C}\mathbb{E}\left[\mathbf{1}\left\{\frac{1}{\pi(i)_y}\bar{V}^\pi(i) - \frac{\pi(i)_x}{\pi(i)_y} \in \cdot\right\}\frac{\pi(i)_y}{\mu_i(V^\pi(i))}\right]$$

where $C = \mathbb{E}\left[\dfrac{\pi(i)_y}{\mu_i(V^\pi(i))}\right]$.

If $0 < (\mathbb{E}\xi)^*(S) < \infty$ the distribution of the typical cell of π may be written as (identifying $\Omega \times \{i\}$ with Ω)

$$\int \mathbf{1}\{\bar{C}^\pi(\omega, i) \in \cdot\}\mathbb{P}^\xi(d\omega),$$

while the Δ-volume weighted typical cell has distribution

$$\frac{1}{\mathbb{E}^\xi \mu_i^\Delta(C^\pi)}\int \mathbf{1}\{\bar{C}^\pi(\omega, i) \in \cdot\}\mu_i^\Delta(C^\pi(\omega, i))\mathbb{P}^\xi(d\omega),$$

whenever $0 < \mathbb{E}^{\xi}\mu_i^{\Delta}(C^{\pi}) < \infty$, where μ_i^{Δ} is given by (7.27). Corollary 7.16 applies to these distributions and gives in particular the relations

$$\mathbb{E}^{\xi}\left[\mu_i^{\Delta}(C^{\pi}(\theta_e, i))\right] = \frac{1}{\gamma_{\xi}},$$

$$\mathbb{E}\left[\frac{\Delta^*(\pi(i))}{\mu_i(V^{\pi}(i))}\right] = \gamma_{\xi}.$$

7.3   Applications of the integrated MTP

Two applications of the integrated versions of the MTP in the form of Theorem 5.5, more precisely in the special form from (5.11), are given in this section. The first result gives an idea on how to approximate Borel sets in an unbiased way, using a G-stationary random partition. The second gives an interpretation of the intensity measure of k-dimensional Hausdorff measure restricted to the k-skeleton of a G-stationary tessellation in $\mathbb{R}^d$.

7.3.1   Unbiased approximation of Borel sets

Consider $\mathbb{R}^d$ with Lebesgue measure λ^d. It is well known that the group of rigid motions G_d on $\mathbb{R}^d$ becomes a locally compact, second-countable Hausdorff group that operates continuously and topologically properly on $\mathbb{R}^d$ when endowed with a suitable topology (see e.g. [64, Theorem 13.2.3]). It is further unimodular as the proof in [64, Theorem 13.2.10] shows. All these properties (except perhaps for the unimodularity) are inherited from G_d by any closed subgroup G of G_d when endowed with the trace topology also operates topologically properly on $\mathbb{R}^d$. We now fix a closed unimodular subgroup G of G_d.

Examples.   (i) G might be a linear subspaces L of $\mathbb{R}^d$ of dimension k where $0 \le k \le d$, that acts on $\mathbb{R}^d$ via translation. Orbits of such an operation are all k-dimensional affine subspaces parallel to L.

(ii) G may be a discrete additive subgroup Γ of $\mathbb{R}^d$ that also act on $\mathbb{R}^d$ via translation. These include the additive subgroups of $\mathbb{R}^d$ generated by finitely many vectors from $\mathbb{R}^d$. The orbits here are the translates of the grid Γ.

(iii) G may equal $SO(d)$ or a lower dimensional rotation group (isomorphic to $SO(k)$ where $0 < k \le d$).

Let ξ denote a G-stationary simple point process on $\mathbb{R}^d$ $\pi : \Omega \times \mathbb{R}^d \to \mathbb{R}^d$ a G-stationary random partition based on ξ. The letter A denotes a G-invariant measurable subset of $\mathbb{R}^d$ while B denotes any fixed G-symmetric subset of $\mathbb{R}^d$. We define

$$C^{\pi}(\omega, B) := \{x \in \mathbb{R}^d : \pi(\omega, x) \in B\}, \tag{7.28}$$

i.e. $C^{\pi}(\omega, B)$ is the union of all cells with center in B. Note that since we did not specify a σ-algebra on the space of all subsets of $\mathbb{R}^d$ it would not make sense to interpret $C^\pi(B)$ as a random set, or to speak of its distribution. Instead of introducing such a σ-algebra, we consider $\lambda^d(A \cap C^\pi(B))$ and note that this is by Fubini's Theorem a random variable since

$$\lambda^d(A \cap C^\pi(\omega, B)) = \int \mathbf{1}\{x \in A, \pi(\omega, x) \in B\}\lambda^d(dx), \quad \omega \in \Omega.$$

Note that in the special case $G = \mathbb{R}^d$, i.e. the completely stationary case, the symmetry condition on B reduces to the condition $\lambda^d(B) > 0$. The following theorem is even new in this special case which extends results of Heveling and Reitzner [26] from homogeneous Poisson-Voronoi tessellations to general stationary partitions.

Theorem 7.21 (approximation of symmetric sets). *Consider the canonical action of a closed unimodular subgroup G of G_d on $\mathbb{R}^d$. Let ξ denote a G-stationary simple point process in $\mathbb{R}^d$ and let π denote a G-stationary partition based on ξ. Further, let A denote a G-invariant measurable subset of $\mathbb{R}^d$ and B denote a G-symmetric subset of $\mathbb{R}^d$. Then*

$$\mathbb{E}\left[\lambda^d(A \cap C^\pi(B))\right] = \lambda^d(A \cap B), \qquad (7.29)$$

where $C^\pi(B)$ is defined as in (7.28).

Proof. Clearly $\eta(\omega, \cdot) := \lambda^d$ is G-stationary and ξ is G-stationary by hypothesis. Defining the evidently G-invariant kernels $\gamma(\omega, s, \cdot) = \lambda^d(\cdot)$ and $\delta(\omega, t, \cdot) = \xi(\omega, \cdot)$ we clearly have $\xi \otimes \gamma = \eta \otimes \delta$ ω-wise such that (5.11) yields for $C := B$

$$\mathbb{E}\iint \mathbf{1}_B(s)m(\theta_e, s, t)\lambda^d(dt)\xi(ds) = \mathbb{E}\iint \mathbf{1}_B(t)m(\theta_e, s, t)\xi(ds)\lambda^d(dt),$$

where m is an arbitrary jointly G-invariant non-negative measurable function. Choosing here the evidently jointly G-invariant

$$m(\omega, s, t) := \mathbf{1}\{t \in A, t \in C^\pi(\omega, s)\}, \quad s, t \in \mathbb{R}^d, \omega \in \Omega,$$

we receive

$$\mathbb{E}\int \mathbf{1}_B(s)\lambda^d(A \cap C^\pi(s))\xi(ds) = \mathbb{E}\iint \mathbf{1}_B(t)\mathbf{1}\{t \in A \cap C^\pi(s)\}\xi(ds)\lambda^d(dt).$$

Here the left-hand side clearly equals the left-hand side of (7.29). And since for any fixed $t \in \mathbb{R}^d$

$$\int \mathbf{1}\{t \in A \cap C^\pi(\omega, s)\}\xi(\omega, ds) = \mathbf{1}\{t \in A\}, \quad \omega \in \Omega,$$

the right-hand side reduces to the right-hand side of (7.29). $\qquad\square$

The theorem may be interpreted as follows. If the G-symmetric set B is unknown, but $\lambda^d(A \cap C^\pi(B))$ is known from some data, then $\lambda^d(A \cap C^\pi(B))$ represents an unbiased estimator for $\lambda^d(A \cap B)$. In the completely stationary case where $A = \mathbb{R}^d$ and B may be chosen arbitrary with $\lambda^d(B) > 0$ (which is clearly not essential) the theorem reduces to

$$\mathbb{E}\left[\lambda^d(C^\pi(B))\right] = \lambda^d(B),$$

and thus gives the information that given an arbitrary Borel subset B of $\mathbb{R}^d$, the 'approximation' $C^\pi(B)$ is in mean of the same size as B. The above theorem naturally comprises e.g. random tessellations in $\mathbb{R}^d$, as these may be seen as special (slightly modified) random partitions. These are the object of interest in the next subsection.

7.3.2   Intensity measure of a random k-skeleton

Let X denote a G-stationary tessellation of $\mathbb{R}^d$, where again G is assumed to be a closed unimodular subgroup of G_d. We may similarly to (7.28) define for any G-symmetric subset $B \subset \mathbb{R}^d$

$$C^k(B) := \bigcup_{F \in \mathcal{F}_k(X):\pi(F)\in B} F. \tag{7.30}$$

Clearly, if X is G-stationary, then all the N_k and M_k are also G-stationary. Given the configuration ω and a point $s \in N_k(\omega)$, there is, since N_k is simple, a unique k-face F with $\pi(\omega, F) = s$ and we may define

$$C_k(\omega, s) = F.$$

Using the Mass-Transport Principle in the form of Theorem 5.5, more precisely in the special form from (5.11), yields the following theorem (also see the examples in the previous subsection).

Theorem 7.22. (intensity measure of M_k) *Consider the canonical action of a closed subgroup G of G_d on $\mathbb{R}^d$. Let X denote a G-stationary random mosaic of $\mathbb{R}^d$ and let π denote a (generalized or not) G-covariant center function such that N_k defined as in (7.1) is a.s. simple. Further, let M_k be defined as in (7.2), let $A \subset \mathbb{R}^d$ be measurable and G-invariant and $B \subset \mathbb{R}^d$ be G-symmetric. Then*

$$\mathbb{E}\left[\mathcal{H}^k(A \cap C^k(B))\right] = (\mathbb{E}M_k)(A \cap B), \tag{7.31}$$

where $C^k(B)$ is defined as in (7.30).

Proof. Putting $\xi := N_k$, $\eta := M_k$, $\gamma(\omega, s, \cdot) := M_k(\omega, \cdot)$ and $\delta(\omega, t, \cdot) = N_k(\omega, \cdot)$ in (5.11) for $C := B$ yields, using the jointly G-invariant $m(\omega, s, t) = \mathbf{1}\{t \in A \cap C_k(\omega, s)\}$, that

$$\mathbb{E} \iint \mathbf{1}_B(s)\mathbf{1}\{t \in A \cap C_k(\omega, s)\}M_k(dt)N_k(ds)$$
$$= \mathbb{E} \iint \mathbf{1}_B(t)\mathbf{1}\{t \in A \cap C_k(\omega, s)\}N_k(ds)M_k(dt).$$

Here the left-hand side equals the left-hand side of (7.31) and the right-hand side equals the right-hand side of (7.31) since for fixed $\omega \in \Omega$ and $M_k(\omega, \cdot)$-a.a. $t \in \mathbb{R}^d$

$$\int \mathbf{1}\{t \in A \cap C_k(\omega, s)\}N_k(ds) = \mathbf{1}\{t \in A\}. \qquad \square$$

Note that for $k = d$ the above theorem gives a version of Theorem 7.21 for tessellations instead of partitions (up to the minor differences concerning the boundaries of the cells the case $k = d$ in Theorem 7.22 is contained in Theorem 7.21). We illustrate this case for two different groups G in Figures 7.4 and 7.5.

7.4   Applications of the ergodic theorems

In this last section we quickly illustrate the use of our results in Chapter 6 by applying the convergence Theorems 6.15 and 6.13.

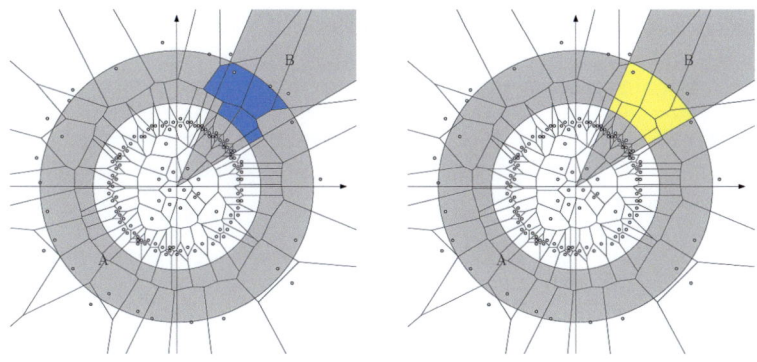

Figure 7.4: $G = SO(2) \hookrightarrow \mathbb{R}^2$: $A \cap C^d(\omega, B)$ on the left and its mean on the right.

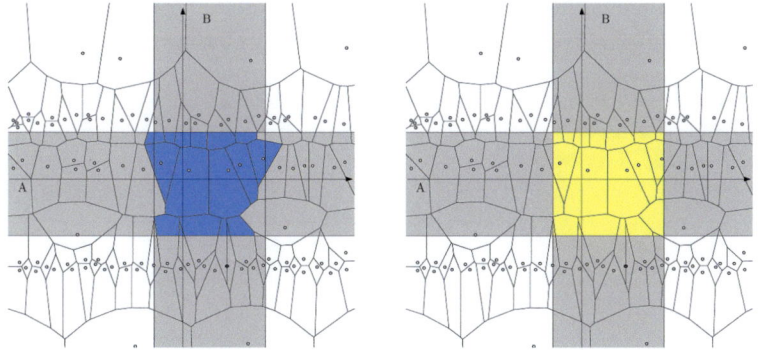

Figure 7.5: $G = \{(x, 0) : x \in \mathbb{R}\} \hookrightarrow \mathbb{R}^2$: $A \cap C^d(\omega, B)$ on the left and its mean on the right.

7.4.1 Grid-stationary case

The following corollary is a consequence of Theorem 6.15.

Corollary 7.23 (grid-ergodic random tessellations). *Let X denote a $\mathbb{Z}^d$-stationary tessellation in $\mathbb{R}^d$ such that for fixed $0 \leq k \leq d$, the point process N_k as defined in (7.1) is $\mathbb{Z}^d$-ergodic. Let A denote a $\mathbb{Z}^d$-invariant set and B_n denote a sequence of $\mathbb{Z}^d$-symmetric subsets of $\mathbb{R}^d$ such that $B_n \cap \mathbb{Z}^d$ is an increasing sequence of boxes. Then, for any measurable and jointly $\mathbb{Z}^d$-invariant $h : \Omega \times \mathcal{C}' \to [0, \infty)$ satisfying*

$$\int_{[0,1)^d \cap A} h(\theta_e, C_k(x)) N_k(dx) \in L \log^{d-1} L(\mathbb{P})$$

it holds that

$$\frac{1}{\delta(B_n)} \int_{B_n \cap A} h(\theta_e, C_k(x)) N_k(dx) \to \int h(\omega, C_k(\omega, b)) \mathbf{1}\{b \in A\} \mathbb{Q}^{N_k}(d(\omega, b)) \quad a.s. \tag{7.32}$$

Proof. We may use the joint $\mathbb{Z}^d$-invariance of h and $\mathbb{Z}^d$-covariance of C_k by writing

$$h(\omega, C_k(\omega, x)) = h(\theta_{x-\beta(x)}^{-1}\omega, C_k(\theta_{x-\beta(x)}^{-1}\omega, \beta(x))), \quad x \in N_k(\omega), \omega \in \Omega.$$

The assertion now follows from Theorem 6.15. $\square$

Examples. Possible choices of h are e.g.

(i) $N_i : \Omega \times \mathcal{C}' \to \mathbb{N}$, where $N_i(\omega, P)$ is defined as the number of adjacent i-faces of the k-face P in configuration ω, if P is a k-face of $X(\omega)$ and is 0 otherwise. This includes the cases $i \leq k$ and $i > k$.

(ii) $V : \mathcal{C}' \to [0, \infty)$ where $V(C) = \mathcal{H}^k(C)$ if C is a k-dimensional polytope and 0 otherwise.

(iii) Even 'arbitrarily big' neighborhoods of the k-faces may be investigates. As an example, fix $n \in \mathbb{N}$ and let $h(\omega, C)$ denote the k-dimensional Hausdorff measure of the union of all k-faces of graph distance at most n from C, if C is a k-face in configuration ω, and 0 otherwise. Here, 'graph distance' refers to the usual graph distance in the k-skeleton of X, interpreted as a graph.

Remark 7.25. Dividing both sides of (7.32) by $(\mathbb{E}N_k)^*(A) = (\mathbb{E}N_k)(A \cap [0,1)^d)$ yields a probabilistic interpretation of the limit. E.g. in two dimensions, we obtain for $A = \mathbb{R}^d$ and if X is such that

$$\int_{[0,1)^d} N_0(C_2(x)) N_2(dx) \in L \log L(\mathbb{P})$$

that

$$\frac{1}{\mathbb{E}N_2(B_n \cap [0,1)^d)} \int_{B_n} N_0(C_2(x)) N_2(dx) \to \int N_0(C_2(\omega, b)) \mathbb{P}^{N_2}(d(\omega, b)) \quad a.s.$$

where the limit may be expressed as

$$n_{20} = \frac{2n_{02}}{n_{02} - 2},$$

see Remark 7.4.

7.4.2   Subspace-stationary case

As similar result to Corollary 7.23 is the following, which is now a consequence of Theorem 6.13.

Corollary 7.26 (subspace-ergodic random tessellations). *Let X denote an L-stationary tessellation in $\mathbb{R}^d$ such that for fixed $0 \leq k \leq d$, the point process N_k as defined in (7.1) is L-ergodic. Let $A \subset \mathbb{R}^d$ denote an admissible L-invariant set and $B_1 \subset B_2 \subset \ldots$ a nested sequence of convex L-symmetric and $L^\perp$-invariant sets in $\mathbb{R}^d$ with $\delta(B_n) \to \infty$. Then, for any measurable and jointly L-invariant $h : \Omega \times \mathcal{C}' \to [0, \infty)$ it holds that*

$$\frac{1}{\delta(B_n)} \int_{B_n \cap A} h(\theta_e, C_k(x)) N_k(dx) \to \int h(\omega, C_k(\omega, b)) \mathbf{1}\{b \in A\} \mathbb{Q}^{N_k}(d(\omega, b)) \quad a.s.$$

$$(7.33)$$

Proof. We may use the joint L-invariance of h and L-covariance of C_k by writing

$$h(\omega, C_k(\omega, x)) = h(\theta^{-1}_{x-\beta(x)}\omega, C_k(\theta^{-1}_{x-\beta(x)}\omega, \beta(x))), \quad x \in N_k(\omega), \omega \in \Omega.$$

The assertion now follows from Theorem 6.17.   □

Remark 7.27. Clearly, the examples given in the previous subsection also apply in this case. A probabilistic interpretation of the limit may be derived here by dividing both sides of (7.33) by $(\mathbb{E}N_k)^*(A)$.

Example 7.28. We consider the infinite cylinder $C = \mathbb{R} \times S^1 \simeq \mathbb{R} \times [0, 2\pi)$ and consider the action of $G = \mathbb{R}$ (with 1-dimensional Lebesgue measure as Haar measure) by shifting the first component. Given an $\mathbb{R}$-stationary tessellation on C, we define as in (7.5) the quantities

$$\gamma^{(i)} := (\mathbb{E}N_i)^*(C)$$

and note that these may be written by means of any G-symmetric set B of width 1 (e.g. $B = [0, 1] \times [0, 2\pi)$) as

$$\gamma^{(i)} := (\mathbb{E}N_i)(B), \quad i \in \{0, 1, 2\}.$$

Exactly as in the proof of Lemma 7.3 one may prove the Euler-type relation

$$\gamma^{(0)} - \gamma^{(1)} + \gamma^{(2)} = 0,$$

and as in Lemma 7.2 it holds that

$$\gamma^{(i)} n_{ij} = \gamma^{(j)} n_{ji}, \quad i, j \in \{0, 1, 2\},$$

where the n_{ij} are defined as in (7.6) and (7.7), now interpreted with respect to the above action. As in Remark 7.4 we conclude that

$$n_{20} = \frac{2n_{02}}{n_{02} - 2}.$$

We may interpret the cylinder as the subset $\mathbb{R} \times [0, 2\pi)$ of $\mathbb{R}^2$. Then clearly $A = C$ is admissible in the sense of the above Corollary and any increasing sequence of $\mathbb{R}$-symmetric sets of the form $B_n = [0, c_n] \times [0, 2\pi)$ with $c_n \to \infty$ may be replaced by the sequence $\tilde{B}_n = [0, c_n] \times \mathbb{R}$, which consists of convex L-symmetric and $L^\perp$-invariant subsets of $\mathbb{R}^2$ on which we may apply the above Corollary 7.26. We obtain

$$\frac{1}{c_n} \int_{B_n} N_0(C_2(x)) N_2(dx) \to n_{20} = \frac{2n_{02}}{n_{02} - 2} \quad a.s.$$

Nomenclature

Bibliography

[1] S. Adams. Trees and amenable equivalence relations. *Ergodic Theory Dynamical Systems*, 10:1–14, 1990.

[2] F. Affentranger. Random spheres in a convex body. *Arch. Math.*, 55:74–81, 1990.

[3] R.V. Ambartzumian. On random plane mosaics. *Soviet Math. Dokl.*, 12:1349–1353, 1971.

[4] R.V. Ambartzumian. Convex Polygons and random tessellations, chapter in: *Stochastic Geometry*, pages 176–191. Wiley, New York, 1974.

[5] V. Baumstark and G. Last. Some distributional results for poisson-voronoi tessellations. *Adv. Appl. Prob.*, 39:16–40, 2007.

[6] I. Benjamini, R. Lyons, Y. Peres, and O. Schramm. Group-invariant percolation on graphs. *Geom. Funct. Anal.*, 9:29–66, 1999.

[7] I. Benjamini and O. Schramm. Percolation in the hyperbolic plane. *J. Amer. Math. Soc.*, 14:487–507, 2001.

[8] B. Bollobás. *Modern Graph Theory*. Springer, New York, 1998.

[9] B. Bollobás and O. Riordan. *Percolation*. Cambridge University Press, 2006.

[10] N. Bourbaki. *Topologie Générale, p.1*. Addison-Wesley, 1966.

[11] L. Breimann. *Probability*. Addison-Wesley, Reading, MA. Repr. SIAM, Philadelphia, 1968.

[12] R. Cowan. The use of the ergodic theorems in random geometry. *Adv. Appl. Prob., Suppl.*, 10:47–57, 1978.

[13] R. Cowan. Properties of ergodic random mosaic processes. *Math. Nachr.*, 97:89–102, 1980.

[14] D.J. Daley and D. Vere-Jones. *An Introduction to the Theory of Point Processes (second edition), vol. I*. Springer, 2003.

[15] D.J. Daley and D. Vere-Jones. *An Introduction to the Theory of Point Processes (second edition), vol. II*. Springer, New York., 2008.

[16] C. Dellacherie and P.-A. Meyer. *Probabilités et Potentiel*. Hermann, Paris, 1975.

[17] R. Diestel. *Graph Theory*. Springer, Berlin Heidelberg, 2006.

[18] R. M. Dudley. *Real Analysis and Probability*. Wadsworth & Brooks/Cole, Belmont, 1989.

[19] J. Elstrodt. *Maß- und Integrationstheorie*. Springer, Berlin Heidelberg, 1996.

[20] G. B. Folland. *Real Analysis - Modern Techniques and their Applications (second edition)*. Wiley Interscience, 1999.

[21] D. Gentner and G. Last. Palm pairs and the general mass-transport principle. *to appear in Math. Z.*, 2010.

[22] G. Grimmett. *Percolation (second edition)*. Springer, Berlin Heidelberg, 1999.

[23] B. Grünbaum. *Convex Polytopes*. Interscience, London, 1967.

[24] O. Häggström. Infinite clusters in dependent automorphism invariant percolation on trees. *Ann. Probab.*, 25:1423–1436, 1997.

[25] S. Helgason. *Differential Geometry, Lie Groups, and Symmetric Spaces*. American Mathematical Society, Providence Rhode Island, 2001.

[26] M. Heveling and M. Reitzner. Poisson-Voronoi approximation. *Ann. Appl. Probab.*, 19:719–736, 2009.

[27] Y. Isokawa. Poisson-voronoi tessellations in three-dimensional hyperbolic spaces. *Adv. Appl. Prob.*, 32:648–662, 2000.

[28] O. Kallenberg. *Foundations of Modern Probability (second edition)*. Springer, New York, 2002.

[29] O. Kallenberg. *Probabilistic symmetries and invariance principles*. Springer, New York, 2005.

[30] O. Kallenberg. Invariant measures and disintegrations with applications to palm and related kernels. *Probab. Th. Rel. Fields*, 139:285–310, 2007.

[31] O. Kallenberg. Invariant palm and related disintegrations via skew factorization. *to appear in Probab. Th. Rel. Fields*, 2010.

[32] O. Kallenberg. Commutativity properties of conditional distributions and palm measures. *Commun. Stoch. Anal.*, 4, 2010, to appear.

[33] O. Kallenberg. Iterated palm conditioning and some slivnyak-type theorems for cox and cluster processes. *to appear in J. Theor. Probab.*, 2011.

[34] J. Kerstan and K. Matthes. Verallgemeinerung eines Satzes von Sliwnjak. *Re. Roum. Math. Pures Appl.*, 9:811–829, 1964.

[35] G. Kummer and K. Matthes. Verallgemeinerung eines Satzes von Sliwnjak, ii-iii. *Rev. Roum. Math. Pures Appl.*, 15:845–870, 1631–1642, 1970.

[36] G. Last. Stationary partitions and palm probabilites. *Adv. Appl. Prob. (SGSA)*, 38:602–620, 2006.

[37] G. Last. Modern random measures: Palm theory and related models, chapter in: *New Perspectives in Stochastic Geometry*, pages 77–110. Clarendon Press, Oxford, 2010.

[38] G. Last. Stationary random measures on homogeneous spaces. *J. Theor. Probab.*, 23:478–497, 2010.

[39] G. Last and H. Thorisson. Invariant transports of stationary random measures and mass-stationarity. *Ann. Probab.*, 37, No.2:790–813, 2009.

[40] J.M. Lee. *Riemannian Manifolds*. Springer, 1991.

[41] J.M. Lee. *Introduction to Smooth Manifolds*. Springer, New York, 2000.

[42] L. Leistritz and M. Zähle. Topological mean value relations for random cell complexes. *Math. Nachr.*, 155:57–72, 1992.

[43] T.M. Liggett. Reversible growth models on symmetric sets. In *Probabilistic Methods in Mathematical Physics, pages 275-301. Academic Press, Boston, MA. Proceedings of the Taniguchi International Symposium held in Katata, June 20-26, 1985, and at Kyoto University, Kyoto, June 27-29, 1985.*, 1987.

[44] R. Lyons and Y. Peres. Probability on trees and networks. Current version available at *http://mypage.iu.edu/~rdlyons/*.

[45] K. Matthes. Stationäre zufällige Punktfolgen. *I. J.-Ber. Deutsch. Math.-Verein*, 66:66–79, 1963.

[46] J. Mecke. Stationäre zufällige Maße auf lokalkompakten Abelschen Gruppen. *Z. Wahrsch. verw. Gebiete*, 9:36–58, 1967.

[47] J. Mecke. Palm methods for stationary random mosaics, chapter in: *Combinatorial Principles in Stochastic Geometry*. Armenian Academy of Sciences Publishing House, Erevan, 1980.

[48] J. Mecke and L. Muche. The poisson voronoi tessellation. I. A basic identity. *Math. Nachr.*, 176:199–208, 1995.

[49] I. Meijering. Interface area, edge length, and number of vertices in crystal aggregates with random nucleation. *Philips Res. Rep.*, 8:270–290, 1953.

[50] R.E. Miles. On the homogeneous planar poisson point process. *Math. Biosci.*, 6:85–127, 1970.

[51] R.E. Miles. The random division of space. *Adv. Appl. Prob., Suppl.*, pages 243–266, 1972.

[52] Miles, R.E. A synopsis of 'Poisson flats in Euclidian spaces', chapter in: *Stochastic Geometry*, pages 202–227. John Wiley, New York, 1974.

[53] J. Møller. Random tessellations in $\mathbb{R}^d$. *Adv. Appl. Prob.*, 21:37–73, 1989.

[54] J. Møller. *Lectures on Random Voronoi Tessellations*. Lecture Notes Statist. **87**, Springer, New York, 1994.

[55] L. Muche. The poisson-voronoi tessellation: relationships for edges. *Adv. Appl. Prob.*, 37:279–296, 2005.

[56] J. Neveu. Processus ponctuels. *École d'Eté de Probabilités de Saint-Flour VI. Lecture Notes in Mathematics*, 598:249–445, 1977.

[57] X.X. Nguyen and H. Zessin. Ergodic theorems for spatial processes. *Z. Wahrsch. verw. Geb.*, 48:133–158, 1979.

[58] A. Okabe, B. Boots, and K. Sugihara. *Spatial Tesselations; Concepts and Applications of Voronoi Diagrams*. Wiley, Chichester, 1992.

[59] B. v. Querenburg. *Mengentheoretische Topologie, 3. Auflage*. Springer, Berlin, Heidelberg, New York, 2001.

[60] W. Rother and M. Zähle. Palm distributions in homogeneous spaces. *Math. Nachr.*, 149:255–263, 1990.

[61] C. Ryll-Nardzewski. Remarks on processes of calls. In *Proceedings of the 4th Berkeley Symposium on Mathematical Statistics and Probability, vol. 2, pp. 455–465.*, 1961.

[62] R. Schneider. *Convex bodies: the Brunn-Minkowski theory*. Cambridge University Press, 1993.

[63] R. Schneider and W. Weil. *Stochastische Geometrie*. B.G. Teubner Stuttgart, Leipzig, 2000.

[64] R. Schneider and W. Weil. *Stochastic and Integral Geometry*. Springer, Berlin, 2008.

[65] I.M. Slivnyak. Some properties of stationary flows of homogeneous random events. *Theory Probab. Appl.*, 7:336–341, 1962.

[66] D. Stoyan, W.S. Kendall, and J. Mecke. *Stochastic Geometry and its Applications (second edition)*. Wiley, Chichester, 1995.

[67] Á. Timàr. Percolation on nonunimodular transitive graphs. *Ann. Probab.*, 34:2344–2364, 2006.

[68] A. Tortrat. Sur les mesures alÁÍatoires dans les groupes non abéliens. *Annales de l'institut Henri Poincaré (B)*, 5:31–47, 1969.

[69] V.I. Trofimov. Automorphism groups of graphs as topological groups. *Math. Notes*, 38:717–720, 1985.

[70] J. van den Berg and R.W.J. Meester. Stability properties of a flow process in graphs. *Random Structures Algorithms*, 2:335–341, 1991.

[71] R. van der Hofstadt. Percolation and random graphs, chapter in: *New perspectives in Stochastic Geometry*, pages 173–247. Clarendon Press, Oxford, 2010.

[72] N. Wiener. The ergodic theorem. *Duke Math. J.*, 5:1–18, 1939.

[73] W. Woess. Topological groups and infinite graphs. *Discrete Mathematics*, 95:373–384, 1991.

[74] A. Zygmund. An individual ergodic theorem for non-commutative transformations. *Acta Scientiarum Mathematicarum (Szeged)*, 14:103–110, 1951.